Axiomatics

Axiomatics

Mathematical Thought and High Modernism

ALMA STEINGART

THE UNIVERSITY OF CHICAGO PRESS CHICAGO AND LONDON

The University of Chicago Press, Chicago 60637
The University of Chicago Press, Ltd., London

Published 2023
Printed in the United States of America

32 31 30 29 28 27 26 25 24 23   1 2 3 4 5

ISBN-13: 978-0-226-82418-5 (cloth)
ISBN-13: 978-0-226-82420-8 (paper)
ISBN-13: 978-0-226-82419-2 (e-book)
DOI: https://doi.org/10.7208/chicago/9780226824192.001.0001

Library of Congress Cataloging-in-Publication Data

Names: Steingart, Alma, author.
Title: Axiomatics : mathematical thought and high modernism / Alma Steingart.
Description: Chicago : The University of Chicago Press, 2023. |
Includes bibliographical references and index.
Identifiers: LCCN 2022026640 | ISBN 9780226824185 (cloth) | ISBN 9780226824208 (paperback) | ISBN 9780226824192 (ebook)
Subjects: LCSH: Mathematics—Philosophy—History—20th century. | Mathematics—United States—History—20th century. | Knowledge, Theory of—United States—History—20th century. | Axiomatic set theory. | BISAC: MATHEMATICS / History & Philosophy | PHILOSOPHY / Epistemology
Classification: LCC QA8.4 .S74 2023 | DDC 510.1—dc23/eng20220829
LC record available at https://lccn.loc.gov/2022026640

♾ This paper meets the requirements of ANSI/NISO Z39.48-1992 (Permanence of Paper).

Contents

Note to Readers

This book is a history of mathematical thought rather than of mathematics per se. The distinction is an important one, not only because my aim is to trace the history of axiomatic thinking outside of the sole confines of mathematics, but also because I have tried to limit to a minimum the mathematical content in the book. The reader does not need to have any prior knowledge of, or even affinity for, mathematics in order to follow the arguments and analysis I make in these pages. The only chapter that asks you to follow a mathematical procedure is chapter 1. Here I felt that outlining some of the basic ideas behind the development of algebraic topology was necessary to appreciate the sea change that occurred when mathematical texts reversed the relation between structure and content in the 1950s and 1960s. The reader who is strongly averse to mathematics can skip the chapter.

Introduction

To see things not abstract as if they were. And vice versa.
There is no abstraction like the present, and the present is entirely concrete. (False.)
Abstraction makes nothing happen.
That's no abstraction, that's my husband.
That's no abstraction, that's the symbolic order writ large.
—Charles Bernstein, "Disfiguring Abstraction"

On the evening of Thursday, December 19, 1946, the great German mathematician Hermann Weyl, an émigré to the United States, pleaded with eighty of the leading lights of modern mathematics while they were trying to eat their dinner. The room before him was packed with mathematicians who had descended upon Princeton University three days earlier for the first international mathematical gathering since the Second World War's end. The tone was meant to be triumphant, a return to intellectual comity after the ravages of war. Now that the fighting was finally over, mathematicians were eager to celebrate their past achievements and announce the start of a new age of research. Those in attendance fell roughly into three groups. The first included old guard American mathematicians like Oswald Veblen and Marston Morse, who had worked tirelessly in the decades before the war to promote research mathematics in the United States and place it on the international stage. The second consisted of European émigrés who poured into the US throughout the 1930s seeking shelter from persecution and the devastation of war (Weyl,

John von Neumann, and Richard Brauer were just a few). The last group was composed of the youngest members of the mathematical community. These young Americans had trained in Princeton, Chicago, and Harvard in the decade before the war, and spent the war itself working for the military and the government on defense-related research. Representing more than a dozen institutions and every subdiscipline, together these eighty men (and they were indeed all men) set the intellectual, social, and pedagogical agendas of American mathematics for the next two decades. Their celebrations, however, came to a halt when the sixty-one-year-old Weyl moved from recording recent mathematical accomplishments to, in his words, "criticizing." Chastising those in the room for their "present indulgence in boundless abstraction," Weyl warned, "I wonder if we have not overplayed the game and in the perpetual tension between the concrete and the abstract have not leaned too heavily on the latter side. The mathematician is in greater danger than the physicist of following the line of least resistance, for he is less controlled by reality."[1]

Twelve years later, the *Atlantic Monthly* published "The Tyranny of Abstract Art," by art historian Ernst Gombrich.[2] A year earlier, Gombrich had published his magisterial *The Story of Art*, which offered an accessible introduction to art history starting in prehistoric times and moving all the way to the "experimental art" of the twentieth century. In the concluding paragraphs of the book, Gombrich claimed the current artistic moment was one of "crisis" in painting and sculpture, whose resolution was yet to be determined.[3] Writing from England, which had become his adoptive home after he fled Austria in 1936, Gombrich now turned his attention to the American audience to criticize what he believed was an overly exaggerated infatuation with abstract art. "The vogue and lure of abstraction is a burning issue in modern art," the article announced. While abstract art had many followers, according to Gombrich, it threatened "to degrade contemporary art to a mere badge of alliances." The critic was wary of the way in which abstraction came to be wed to progress: "Put an abstract into your room and you have proclaimed your allegiance to the right kind of thing, to the future, whatever that may mean."[4]

It was not just in mathematics and art that abstraction was becoming a dominant feature of intellectual and creative work. Ten years after mathematicians gathered at Princeton, it was time for American social scientists to take stock of their own field. A symposium at the University of Chicago on the "State of the Social Sciences" similarly brought together leading figures from political science, economics, sociology, history, and psychology.[5] In

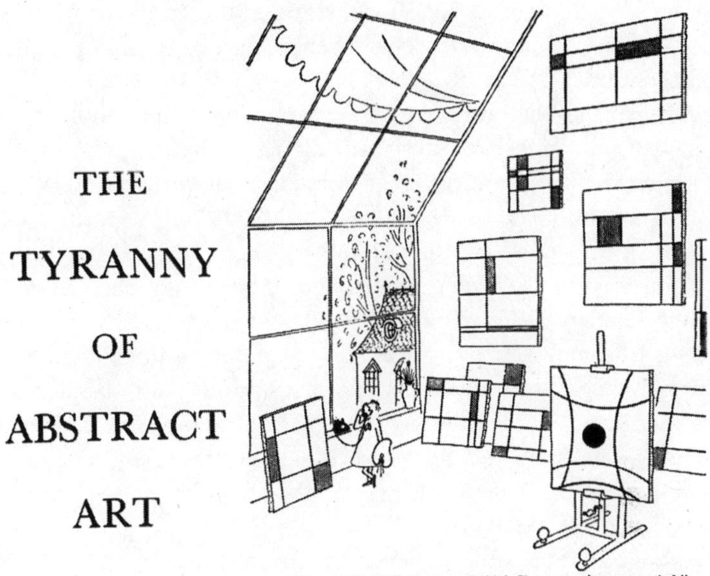

THE TYRANNY OF ABSTRACT ART

"McGraw Gallery? Dolan here. I think I've entered a new period."

BY ERNST GOMBRICH

The vogue and lure of abstraction is a burning issue in modern art. For many painters and critics, faith in abstract art is faith in the future of mankind. But, having become a faith for a few, it has also become a fashion for many, and a fashion which threatens to degrade contemporary art to a mere badge of allegiance. Born in Vienna, Ernst Gombrich *has been a member of the Warburg Institute in London since 1936.*

FIGURE I.1. Ernst Gombrich's article in the *Atlantic* titled "The Tyranny of Abstract Art."
Source: *Atlantic Monthly* 201, 1958

one of the last talks of the event, Friedrich A. Hayek delivered a speech on the "dilemma of specialization," insisting that social scientists should not follow natural scientists in their quest for general laws, but keep their interests "in the particular, individual, and unique event."[6] An Austrian émigré like Gombrich, Hayek had joined the University of Chicago five years earlier, becoming a professor in the Committee on Social Thought. Hayek did not object to theoretical studies out of hand, but he rejected scientism as the model for social sciences:[7] "Not only is the individual concrete instance much more important to us than it is in the natural sciences, but the way from the theoretical construction to the explanation of the particular is also much longer."[8] Yet abstraction was on the rise in the social sciences as well. As Joel Isaac writes, in the postwar period "the possession of systematic conceptual schemes became the *sine qua non* of cutting-edge social science."[9]

What explains the enthusiasm for abstraction at midcentury? Why, despite obvious opposition, did abstraction and theoretical knowledge flourish across such a divergent set of intellectual pursuits? Here, I believe, it is necessary to turn to mathematics. Despite Weyl's heartfelt warnings, those in attendance did not heed his cry. Their faith in abstraction did not diminish in the following decades. On the contrary, abstraction became the defining feature of mathematical thought at midcentury. In 1965, Salomon Bochner, who, like Weyl, had emigrated to the United States in the 1930s, described mathematics as a "repository for abstraction."[10] More than those working in any other field of knowledge, mathematicians pushed abstraction to its most extreme, treating it as both a subject and method. In mathematics, the modernist commitment to abstraction reached its logical conclusion.

There are other reasons to center mathematics in explaining the rise of abstraction. First, for mathematicians, the humanities and the sciences did not compete with one another, but rather were two complementary sides of the same underlying epistemological commitment. Mathematical theories, practitioners argued, were propelled by the internal logic of the field, with no necessary appeal to external reality. They were free creations of the human mind. This abstractionist conception of the field was not limited to pure mathematics. The postwar decades witnessed an unprecedented expansion not only in this arena but also in the mathematization of other academic fields. Mathematicians credited the growth of these new mathematical pursuits—information theory, operations research, numerical analysis, and others—to the abstract nature of their work. For example, Bochner explained, "There is a growing demand for abstraction from possibility in ever wider areas of organized knowledge, especially in the natural sciences and social sciences. It is this which makes many sciences seek affiliation with mathematics; they want abstraction of this kind, much more than the mathematical techniques themselves."[11] The abstract nature of mathematics undergirded both its purity and its utility.

This duality partially explains why abstraction thrived within the period's volatile politics of knowledge. It was the ideal remedy for Cold War anxieties, functioning as a uniquely useful and generative logical contradiction; something that is abstract divorces itself from the lived world, yet abstract knowledge is also primed to be generalizable and therefore applicable in myriad domains. Practitioners of abstraction, from mathematics to the social sciences, could turn away from the vagaries of social and political disaster and declare themselves divorced from the "real." Yet they could also claim that their work furthered the cause of human development by being germane to, and even necessary for, the technologies and theories that came to define American life. This contradiction proved capacious, allowing thinkers and actors epistemic room to maneuver creatively while denying any ideological stance.

Finally, attention to the world of midcentury mathematicians is necessary to appreciate an often-neglected feature of postwar intellectual thought: the axiomatic method. If abstraction was the theory, then axiomatics was the method. Modern axiomatics emerged out of a turn-of-the-century foundational debate in mathematics (which will be discussed in greater detail later), but by the 1950s mathematicians had taken it on as their dominant research tool. In his 1946 speech, Weyl warned those in attendance not to pursue axiomatics at the expense of other methods; it was associated with the most abstruse tendencies of pure mathematicians in the mid-twentieth century. However, as they emerged from the war, those in the field also looked to axiomatics to underline some of the most pragmatic mathematical research of the time. Axiomatic thinking extended first to physical theories such as quantum mechanics.[12] But in the postwar period, the greatest influence of axiomatics outside of mathematics was not in physics but in the social sciences and a host of new disciplines that emerged from the war.

The axiomatic method fundamentally reconfigured the meaning of mathematization at a time when more and more disciplines were turning to mathematics. Following Theodore Porter, historians understand the midcentury mathematization of the social sciences almost exclusively in terms of *quantification*, whereby "trust in numbers" was commensurate with objectivity.[13] However, when postwar social scientists annexed mathematics for use in their theories of social behavior, they did so not by quantifying the phenomena they sought to explicate but by *axiomatizing* them.[14] In other words, it was not quantification per se that rendered the social sciences amenable to mathematization, nor a belief in the objective nature of numbers, but, more fundamentally, a faith in abstraction as a central methodology. The application of mathematics entailed adopting a governing philosophy that conceived of knowledge as universalized and generalized theories. It also privileged the formal consistency of a theory above its phenomenological correspondence, and held that decontextualized analysis was necessary for revealing the structural essence hidden in the accumulation of details.

Axiomatics focuses on the worlds of midcentury American mathematicians during this time of great transformation. It follows a series of controversies that erupted among mathematicians about the nature of mathematics as a field of study and as a body of knowledge. As the field was expanding in both scale and scope, no clear agreement existed among mathematicians on the seemingly simple question of what mathematics *was*. The ensuing debates offer a window not only into the postwar development of mathematics but also into Cold War epistemology writ large. Far from being divorced from the rest of the world, mathematicians had to actively and transparently demonstrate the ways in which knowledge was produced, argued about, and adjudicated. As such, mathematics allows us to make better sense of modes of logic, proclamations of truth, and notions of objectivity that would otherwise go uncontested in other locales. It is the social activity par excellence in which styles of truth—here, abstraction—become synonymous with ways of knowing.

Abstraction and Modernism

But what is abstraction? And how does the mid-twentieth-century fascination with it fit into its longer history? Here I outline two answers, one following abstraction's history as a mode of thought and the other as a marker of modernism.[15] Taken as a mode of thought, abstraction harkens back to the Greek *aphaeresis*, meaning "to take away." Aristotle confined his use of *aphaeresis* to describing the formation of mathematical objects or concepts, but it later expanded to denote the formation of universal concepts more generally.[16] The medieval philosopher Boethius introduced the word *abstraction* into Latin in the sixth century. In his commentary on Porphyry's *Isagoge*, Boethius appeals to abstraction to support a theory of "moderate realism." Refusing to acknowledge the separate reality of universals, Boethius nonetheless holds that, through the act of abstraction in thought, concepts possess a strong basis in reality: "But the understanding that does this by division and abstraction and taking things away from the things they exist in, not only is not false, but is almost able to find out what is properly true."[17] That is, for Boethius, abstraction functioned as an intellectual operation, gathering the likenesses of qualities existing across objects into universals. It is with this passage that the word "abstraction" enters the pit of philosophical debate.

The problem of universals was a hallmark of medieval philosophy, attracting the attention of Thomas Aquinas and William of Ockham, among many others. Do universals, such as the color red, the idea of beauty, and the geometrical line, exist in reality, or are they merely figures of speech contained by thoughts? The debate revolved around the ontological status of universals, not the intellectual operation involved in the practices of abstraction that created them.[18] That frame shifted in the eighteenth century, when George Berkeley mounted a strong critique of abstraction as a mode of thought, and of general abstract ideas.[19] A staunch idealist, Berkeley published on metaphysics, theology, optics, and the philosophy of mathematics. One of his earliest works, written when he was in his mid-twenties, was *A Treatise Concerning the Principles of Human Knowledge* (1710), which strongly rebukes abstraction.[20] Berkeley takes aim at the belief that the mind can consider qualities such as color, shape, or motion in the abstract. We are told, Berkeley writes, that the mind, "by leaving out of the particular colors perceived by sense, that which distinguishes them one from another, and retaining that only which is common to all, makes an idea of color in abstract which is neither red, nor blue, nor white, nor any other determinate color."[21] But is that even possible? Can one really think about the idea of color without immediately thinking of a particular color? For Berkeley, the answer to these questions was a resounding no. We do not possess some general abstract idea of color as such.

Berkeley did not, however, dismiss the mind's ability to abstract in toto. As he explained, one can abstract in one's mind the idea of an eye, a head, or a leg, because one can imagine these ideas separately. However, the same is not true of general abstract ideas, such as color or motion, since no matter how hard you try, you cannot picture these ideas in the abstract. Berkeley's criticism was directed, at least in part, toward John Locke, who in his *Essay Concerning Human Understanding* (1689) claimed that we are able to equally apply the word "triangle" to isosceles, right, obtuse, and equilateral triangles because we possess an abstract idea of a triangle to which all these examples refer. The general idea of a triangle, Locke writes, is "neither oblique, nor rectangle, neither equilateral, equicrural, nor scalenon; *but all and none of these at once*."[22] Berkeley would have none of that, and argued instead that "there is no such thing as one precise and definite signification annexed to a general name, they all signifying indifferently a great number of particular ideas."[23] The word *triangle* signifies a multitude of particular triangles, not some abstract idea of one. Berkeley held that the problem was, at least in part, linguistic. Words, he wrote, produced the "doctrine of abstract ideas." Scholars have since argued that Berkeley mischaracterized Locke's position, but his influence is unmistakable.[24] In his *Treatise of Human Nature* (1739–40), David Hume describes Berkeley's conclusion as "one of the greatest and most valuable discoveries that has been made of late years in the republic of letters."[25] He then offers his own analysis of how abstract ideas form.

Hume argues that, having recognized a certain resemblance among various objects, we attach a word to it despite differences among the particularities; when we use that word, we do not bring to mind all the possible particularities to which the word might refer. The use of the word *tree*, for example, does not require that we simultaneously hold in our minds all kinds of trees—chestnuts, oaks, evergreens, firs, etc.—because the possibilities are practically infinite. Nor, as Berkeley made clear, does "tree" denote some abstract idea of a tree, shorn of all identifying features. Hume writes, "The word not being able to revive the idea of all these individuals, but only touches the soul, if I may be allowed so to speak, and revives the custom, which we have acquired by surveying them."[26] The word, according to Hume, raises an individual, not a general, idea, but it also brings to mind the custom—the process—by which we acquire it. Thus we are able to access any other individual idea the word applies to. What Locke, Berkeley, and Hume made clear is that the problem of abstraction as an operative mode of thought touches directly on the foundations of geometry, the signification function of language, and the relation between perception and reason. All three aspects would continue to trouble midcentury scholars, who were similarly concerned about the role of abstraction in a theory of knowledge. But apprehension about abstraction reached beyond philosophy *strictu sensu*.

Here the work of Edmund Burke is exemplary. Known for both his philosophical and political writing, Burke studied Locke and (probably) Berkeley, and early in his career showed similar interests in the constitution of ideas. Burke dedicated the final chapter of *A Philosophical Enquiry into the Origin of Our Ideas of the Sublime and the Beautiful* (1757) to the subject of words and their ability to excite passions and emotions. Following Berkeley, Burke argues that the ability of language to arouse passion is not an outcome of an abstract idea but, rather, is rooted in the associative power of speech. Richard Bourke explains, "According to Burke, a word like 'honour' receives its meaning by repeated use on appropriate occasions."[27] However, Burke's mistrust of abstract ideas is most apparent not in his philosophical work but in his political writings. In *Reflections on the Revolution in France* (1790), Burke assails those who approach politics in terms of abstract principles.[28] "But I cannot stand forward, and give praise or blame to anything which relates to human actions and human concerns on a simple view of the object, as it stands stripped of every relation, in all the nakedness and solitude of metaphysical abstraction." Burke insists that when it comes to political affairs, experience rather than philosophy should guide one's actions.

Most famously, Burke applies this criticism to assail the idea of abstract rights as articulated in the French Declaration of the Rights of Man in 1789. Philosophers, according to Burke, could spend their days debating the ideal rights of man, but when it came to deciding practical affairs, their musings were worse than useless.[29] Instead of discussing a "man's abstract right to food or medicine," Burke writes, it would be better to come up with a plan to procure and administer the two: "In that deliberation I shall always advise to call in the aid of the farmer and the physician, rather than the professor of metaphysics."[30] A government, he insists, is not based on such perfectly abstract ideas of natural rights. Indeed, "their abstract perfection is their practical defect."[31] Burke reasons that in a civil society, it is necessary to have some restrictions on individual freedom, and since the real rights of man are not permanent but historically and contextually contingent, "they cannot be settled upon any abstract rule; and nothing is so foolish as to discuss them upon that principle."[32] The problem with abstract ideas such as "universal rights," he says, is not philosophical but political. It causes real damage. As Burke saw it, the French people, in the name of some abstract naive idea, had destroyed their country: "They have found their punishment in their success. Laws overturned; tribunals subverted; industry without vigour; commerce expiring; the revenue unpaid, yet the people improvised."[33] Burke here is less interested in the formation of abstract ideas and more in their influence on practical affairs.

Burke, of course, was not alone. Many thinkers have considered the harm done by abstraction in social and political life, perhaps most notable among them Karl Marx. A vast literature exists on both the importance of abstraction to Marx's dialectics and on Marx's critique of abstraction as sustaining the capitalist mode of production (as, for example, when he writes that "individuals are now ruled by abstractions").[34] One such abstraction, for Marx, was general labor, which he attributes to Adam Smith:

> This abstraction of labour as such is not merely the mental product of a concrete totality of labours. Indifference towards specific labours corresponds to a

> form of society in which individuals can with ease transfer from one labour to another, and where the specific kind is a matter of chance for them, hence of indifference. Not only the category, labour, but labour in reality has here become the means of creating wealth in general, and has ceased to be organically linked with particular individuals in any specific form.[35]

According to Marx, in the most developed and "most modern form of existence of bourgeois society"—namely the United States—the abstraction of the category of labor ("labour as such")[36] becomes embedded in material practices. Only in a bourgeois society does the category of labor—not agricultural or manufacturing labor but labor in general, as a wealth-creating activity—emerge. I point to Burke and Marx because, beyond the philosophical problem of abstraction, what they drew attention to were questions that occupied many mid-twentieth century thinkers: namely, what is the "proper" level of abstraction in social or political theory? And what happens when abstract ideas become concrete reality?

To this long history of abstraction as a mode of thought one can add a much shorter one, which weds abstraction to modernism. In *A Singular Modernity*, Fredric Jameson remarks on the frustration that accompanies any analysis of modernism: "Like the pane of glass at which you try to gaze even as you are looking through it, you must simultaneously affirm the existence of the object while denying the relevance of the term that designates that existence."[37] Despite modernism's problematic definition, its porous boundaries and leaky chronology, abstraction remains a relatively stable feature of the philosophical and cultural movement that transformed Western thought during the first few decades of the twentieth century. Yve-Alain Bois goes so far as to suggest that "one could say that it is abstraction that, both retrospectively and programmatically, established modernism as a whole as an enterprise of motivation."[38]

I identify this as another origin story of abstraction because the association of abstraction and modernism is *not* a continuation of the long philosophical and political tradition I just outlined, but instead represents a clean break from it. In announcing a break with the past, in turning against the long history of thinking about abstraction as an operational mode of thought ("to draw away"), abstractions came to be identified with modernism. In philosophy, mathematics, physics, and the arts, abstraction became by midcentury emblematic of a new paradigm of intellectual and artistic creativity, announcing a rupture (whether real or imagined) with the work of the past. In particular, abstraction became modern when it was defined in opposition to its etymological root, meaning to *draw away*. Turn-of-the-century abstraction in mathematics, philosophy, and the arts did not emanate from the world; it was meant to operate independently of it.

This severing of abstraction from the phenomenological world is evident in the arts. In 1915, Kazimir Malevich exhibited his first *Black Square* painting and published his manifesto, *From Cubism and Futurism to Suprematism*. As he later explained, "In the year 1913 in my desperate attempt to free art from the ballast of objectivity, I took refuge in the square form."[39] The freedom Malevich describes was achieved not by looking to the world but away from it. That this new meaning of abstraction was novel was indicated by a 2012 MoMA retrospective entitled *Inventing Abstraction 1910–1925*. The idea that abstraction was "invented" in the first decades of the twentieth century only makes sense if we understand modernist abstraction not as an operative mode of thought but as a rejection of that tradition. Leah Dickerman, the exhibit's curator, writes that "before 1911 . . . it seems to have been impossible for artists to step away from a long-held tenet of artistic practice: that paintings describe things in a real or imaginary world."[40] What, then, made modern abstraction possible? Dickerman points to new physiological theories on the nature of perception, changes in modern scientific thought, Ferdinand de Saussure's semiotic theory of language, the rise of new media culture (the telegraph, the gramophone, photography, the cinema), and Cubism as some of the historical conditions that enabled modernist abstraction to emerge.[41]

Paintings, of course, have always been abstracted, at least in the etymological sense: a Paul Cézanne painting might be representational, but its object was abstracted from the world. Historian of art Hubert Damisch sought to emphasize this point in relation to another (planned yet failed) centenary exhibition on abstract art, this time at the Centre Pompidou in Paris. Seeking to narrate a long history of abstraction, Damisch turned to ancient Greece and the Middle Ages to emphasize a *longue durée* continuity of artistic abstracting. In an interview in 2005, Damisch explained:

> We would have developed the idea that, in a way, abstraction always formed part of the general process of Western art, and that some sort of a crisis developed at the beginning of the twentieth century, to be paralleled with the crisis in mathematics that was looking for its own fundamentals. In the same way that painting was looking for its own specificity, its own fundamentals. Due to such a crisis—the crisis of abstraction, abstraction as a crisis—abstraction was turned into the central issue of art.[42]

Damisch is correct in pointing to the parallel between mathematics and art at the beginning of the twentieth century. As in art, modernist mathematics has been identified with abstraction.[43] And as in art, mathematics became abstract when mathematicians no longer believed that their mathematical concepts were drawn or abstracted from the world.

Art and mathematics might be the clearest examples of abstraction's reversal of meaning at the turn of the twentieth century, but the phenomenon was much broader, impacting a wide spectrum of turn-of-the-century intellectual activity. Above all, modernist abstraction points to a *crisis of referentiality*, one similarly felt in linguistic and scientific philosophy. In linguistics, the work of Ferdinand de Saussure is the clearest example. Saussure's structural linguistics transformed reference from a problem of mimesis to one of signification. He famously distinguished between two aspects of a sign, the "signifier" and the "signified," the former corresponding to a string of sounds or written marks and the latter being the concept or psychological impression. As we have seen with Locke and Hume, the problem of abstraction was entangled with that of the signification function of language. But whereas they turned to abstraction to explain how a word and a concept relate to one another, Saussure asserted that the relation between the signified and signifier was random: "The bond between the signifier and the signified is arbitrary. Since I mean by sign the whole that results from the associating of the signifier with the signified, I can simply say: *the linguistic sign is arbitrary*."[44] In the following decades, Saussure's ideas were developed by the Prague linguists and popularized in the United States by Roman Jakobson. Their influence would reach far beyond linguistic theory.

To a certain degree, Einstein's theories of relativity similarly spoke to a crisis of referentiality. One of the features that distinguishes Einstein's work from that of his predecessors is the assertion that the basic concepts of a physical theory are arbitrary, "a free creation of the human mind," and do not necessarily need to be abstracted from reality.[45] As he explains in his autobiography, "A theory can be tested by experience, but there is no way from experience to the setting up of a theory."[46] That physicists and philosophers alike conceived of the new theory of relativity in terms of abstraction is clearly evident in the writings of Alfred North Whitehead. In *Science and the Modern World*, Whitehead proclaimed that science had reached a turning point: "The old foundations of scientific thought are becoming unintelligible. Time, space, matter, material, ether, electricity, mechanism, organism, configuration, structure, pattern, function, all require reinterpretation."[47] In conclusion he noted, "Thought is abstract; and the intolerant use of abstraction is the major vice of the intellect."[48] It was the job of the philosopher to clear up the uses and abuses of abstraction.

By the middle of the twentieth century, the implications of this crisis of reference were felt across intellectual and artistic activity, and a few dominant theories emerged in reaction. Across various fields, the old theory of abstraction gave way to analytic realism and formal systems. Structuralism moved from linguistics into the human sciences, the biological sciences, and philosophy. In the arts, formalism emerged in both literary and art criticism, and in scientific philosophy the semantic tradition became dominant. In this book I follow another response to this crisis: axiomatics, which has so far received much less attention than the others. Axiomatics emerged in mathematics but, as I make clear, its influence ranged much more widely, to the social sciences, architecture, and philosophy.

Axiomatics

Like abstraction, modern axiomatics underwent an etymological reversal at the turn of the century, when mathematicians shifted from conceiving of axioms as self-evident truths to taking them to be completely arbitrary propositions. Indeed, in mathematics, the two reversals depended on one another. Responsible for the change was German mathematician David Hilbert. In 1899, Hilbert published *The Foundations of Geometry*, a book based on a series of lectures he had given the previous year. The work emerged from mathematicians' and philosophers' general concern for the foundation of mathematics at the end of the nineteenth century. Despite centuries of mathematical research, mathematicians did not turn their attention to the logical foundations of the real number system until this period.[49] A century-old campaign to rigorize analysis boiled down to a seemingly simple question: What are numbers? Within the new regime of objectivity, neither the empiricist nor the idealist answer was acceptable. Numbers were not arrived at by counting, nor were they synthetic a priori.[50] Moreover, the discovery of space-filling curves and other counterintuitive mathematical curiosities made mathematicians suspicious of intuition, even (or especially) when it came to the most elementary mathematical concepts. They associated intuition not only with subjectivity but also with lack of rigor.[51]

An astonishing number of authors felt compelled to write about the question of numbers. Philosophers such as Hermann von Helmholtz and

Edmund Husserl, mathematicians such as Richard Dedekind and Giuseppe Peano, and logicians Gottlob Frege and Bertrand Russell were just a few of the many thinkers who tackled the issue at the turn of the century.[52] Their investigations were not limited to the foundation of mathematics, nor to the desire to rigorize analysis. Rather, numbers represented a special case—a limit case—for the elementary constitution of knowledge. It was the ultimate epistemological query. And the multitude of possible responses to the question reflected changing ideas about objectivity, truth, and certainty.

From the perspective of midcentury American mathematicians, Hilbert's work had the greatest influence.[53] Hilbert began his 1899 book by posing a set of elements (points, lines, and planes) and axioms describing the relations between them (e.g., between any two points there exists one line). Yet here is where terminology can get confusing. Unlike his predecessors, Hilbert did not seek to establish the soundness of his theory from the bottom up. Truth was a property of the system, not of its individual constituents. While he used the words "points" and "lines" to name the elements of his system, they served as empty signifiers that had no direct relation to their intuitive meaning (they were not arrived at by abstraction). Instead, for Hilbert the elements of the system were defined implicitly as those things that satisfied the given set of axioms. This is why Hilbert reportedly remarked that, instead of points, lines, and planes, one should be able to speak freely of tables, chairs, and mugs.

In Hilbert's hands, the word "axiom" no longer denoted a self-evident truth. Rather, the choice of axioms was arbitrary. Their soundness, their accuracy, was not a discrete property but dependent on the system as a whole. Hilbert argued that a mathematical theory was valid as long as its axioms were independent (i.e., neither axiom could be derived from the others) and the system as a whole was consistent (i.e., you could not deduce a statement and its contradiction from the set of axioms). As long as his system was able to reproduce all the known theorems of Euclidean geometry, then it was, by Hilbert's standards, a success.

To appreciate the radical nature of Hilbert's idea, it is useful to examine Gottlob Frege's contemporaneous reaction to the publication.[54] Having read the book, Frege initiated a correspondence with Hilbert. "I think it is about time," he wrote in his first letter on December 27, 1899, "that we came to an understanding about what a definition is supposed to be and do." He then added, "It seems to me that complete anarchy and subjective caprice now prevail."[55] Frege first took issue with Hilbert's insistence that elements of a theory are to be defined implicitly. The "axioms and theorems can never try to lay down the meaning of a sign or word that occurs in them, but it must already be laid down," he wrote.[56] Unless the basic constituents of a theory were clearly defined, Frege objected, how could its statements make any sense? Hilbert thought otherwise. "I do not want to presuppose everything as known," he wrote to Frege. "If one is looking for other definitions of point, perhaps by means of paraphrase in terms of extensions, etc., then, of course, I would most decidedly have to oppose such an enterprise. One is then looking for something that can never be found, for there is nothing there, and everything gets lost, becomes confused and vague, and degenerates into a game of hide-and-seek."[57] For Hilbert, the failure to come up with a satisfying definition of a point was not a testimony to the fact that all past attempts were insufficient, but rather that the project itself was fruitless. If mathematicians were seeking to secure sound foundations for their theories, they had to shift their emphasis away from concepts.

Yet Frege's criticism did not end there. He protested Hilbert's contention that the axioms were arbitrary. For him the axioms of geometry were consistent simply because they accorded with physical reality. Their consistency was guaranteed by their veracity. Hilbert could not disagree more. "For as long as I have been thinking, writing, lecturing about these things, I have been saying the exact reverse: If the arbitrary given axioms do not contradict one another, then they are true, and the things defined by the axiom exist. This is for me the criterion for *truth* and *existence*."[58] In his last letter to Frege, Hilbert repeated his conviction that the only way to define a concept was through its relations to other concepts. "I did not think up this view because I had nothing better to do," he concluded, "but I found myself forced into it by the requirements of strictness in logical inference and in the logical construction of a theory."[59] The two agreed to disagree.

Frege's criticism aside, Hilbert's book was an immediate success and garnered the attention of the entire mathematical community.[60] Reflecting upon the influence of the book almost a hundred years after its publication, Garrett Birkhoff exclaimed that the publication was "without question . . . the most influential book on geometry written in the past century."[61] Yet, even this statement insufficiently describes the book's impact, which extended far beyond geometry. For it was not strictly within the philosophy of mathematics that Hilbert's formal analysis had its greatest impact on the history of abstraction. Throughout the first few decades of the twentieth century, mathematicians adopted Hilbert's methodology and extended it beyond the foundations of mathematics. Especially in the United States, axiomatic analysis came to dominate mathematical research, and in the process further divided abstraction from the concrete and the physical.[62]

In December 1902, Eliakim Hastings Moore gave a presidential address before the American Mathematical Society (AMS). [63] He began by defining abstract mathematics: "The notion within a given domain of defining the objects of consideration rather by a body of properties than by particular expressions or intuitions is as old as mathematics itself. And yet the central importance of the notion appeared only during the last century—in a host of researches on special theories and on the foundations of geometry and analysis. Thus has arisen the general point of view of what may be called abstract mathematics."[64] The identification of abstract mathematics with the study of axiomatic systems became prominent in the United States. For many young mathematicians who joined the field during the first few decades of the twentieth century, Moore's definition was not a philosophical rumination but a research agenda, one to which they referred as "the abstract point of view."

To a certain degree, this book could be read as a story of how Hilbert's modern axiomatics colonized twentieth-century American thought. In the 1940s and 1950s, pure mathematicians adopted the approach and extended it far beyond Hilbert's original intent. Axiomatics became the de facto mathematical method, giving rise to the abstractionist theory of knowledge that dominated midcentury mathematical thought (chapter 1). Modern axiomatics also had a lasting influence on the growth of applied mathematics in the United States by redefining the relation between theory and phenomenon and mathematicians' understanding of what mathematization entails (chapter 2). Hilbert had advocated the use of modern axiomatics in elucidating social scientific theories as early as 1917. In a lecture entitled "Axiomatic Thought," Hilbert argued that theoretical accuracy in every discipline depended on axiomatics: "I believe anything at all that can be the object of scientific thought becomes dependent on the axiomatic method, and thereby on mathematics, as soon as it is ripe for the formation of a theory."[65] As if fulfilling Hilbert's prognosis, in the postwar period American social scientists began adopting axiomatic thinking in their theories of social behavior (chapter 3). Further, mathematicians' insistence that the axioms of a system were by definition arbitrary, by midcentury, testified to the creative power of their research and its subsequent claim to the humanities (chapter 4).

High Modernism

The transformation and adaptation of the axiomatic method in the following decades defined high modernist mathematics as distinct from the period that historians of mathematics have termed the modernist transformation in mathematics (roughly 1890–1930).[66] Starting in the 1930s and expanding in the postwar period, axiomatics came to be associated with universalizing theories and an emphasis on structure. Integral to this move was the work of Nicolas Bourbaki, a pseudonym for a group of French mathematicians who took it upon themselves to rewrite French mathematical textbooks in a modern style. As Bourbaki explained in its most programmatic English-language article, which was published in 1950 in the *American Mathematical Monthly*, "In a single view, it [mathematics] sweeps over immense domains, now unified by the axiomatic method, but which were formerly in a complete chaotic state."[67] Bourbaki held that mathematics consists of a hierarchy of structures, and that these abstract objects were the proper subject of mathematical research.[68]

However, high modernism was not restricted to Bourbaki and should not be equated with it. The influence of the group in the United States was unmistakable, and many of its members spent parts of their careers in American universities.[69] While the group is the most extreme example of mathematical high modernism, many midcentury American mathematicians also adhered to universal theories, axiomatics, and formal presentation, and enforced these principles through the postwar educational system. Even those mathematicians who did not accord with Bourbaki's grand theory of mathematical structures would often share at least some of its methodological interests, especially axiomatics. Moreover, what distinguished postwar high modernism was that mathematicians pointed to the axiomatic method as responsible for both the most esoteric theories of pure mathematics *and* its growing applicability.

Thus, in attaching "high modernism" to midcentury American mathematics, I explicitly call upon two existing scholarly traditions I wish to hold in tandem throughout the book. The first hearkens back to the work of James Scott, who identified high modernism as a belief in the potential of science and technology to order and regulate the social world. "*High* modernism," he wrote, "is thus a particularly sweeping vision of how the benefits of technical and scientific progress might be applied—usually through the state—in every field of human activity."[70] More recently, Hunter

Heyck, following Scott, has proposed the notion of "high modernist social science" to describe the research approach that dominated the American social sciences in the 1950s and 1960s. According to Heyck, midcentury practitioners such as Herbert Simon, Talcott Parsons, David Easton, and Paul Samuelson subscribed to a "bureaucratic worldview" that envisaged the "world as a set of complex, hierarchic systems (a treelike structure)."[71] Every social phenomenon, in this view, could schematically be reduced to a set of relations, mechanisms, and processes. Working in new institutional settings and answering to new funding agencies, social scientists strove to distinguish themselves from the humanities and align their work with the sciences.

A separate tradition exists in the arts, where high or late modernism describes a different group of elite intellectuals.[72] According to Robert Genter, critics such as Theodor Adorno, Clement Greenberg, John Ransom, and Lionel Trilling, despite their many differences, were united in arguing for the place of art in modern American society: "As federal funds flowed into universities from military research and scientists became national celebrities, high modernists waged a campaign within universities, publishing houses, and art galleries to defend the humanities as antidote to modern science."[73] These ideologists of modernism, to borrow Fredric Jameson's language, were united above all in their adherence to the autonomy of the aesthetic. "By purging it [art] of its extrinsic elements, such as the sociological or the political; by reclaiming aesthetic purity from the morass of real life, of business and money, and bourgeois daily life all around it," late modernists sought to reclaim artistic production as a sacred domain.[74] Art offered a corrective to the dominant bureaucratic worldview and the encroachment of scientific thinking into every domain of social life. As such, the goal of cultural and literary criticism was to define and defend the intellectual sphere of educated Americans.[75]

These seemingly contradictory accounts of high modernism, denoting the rationalization of social life on the one hand and the promotion of an autonomous aesthetic sphere on the other, are especially relevant to any discussion of the postwar reconstruction of mathematics. This was, to say the least, a curious time for mathematicians. The war had ignited interest in applied mathematics in the United States. The field, which had been in an almost complete state of stagnation in the prewar decades, suddenly received the attention and lavish endorsement of funding agencies. Game theory, communication theory, operations research, and computing all began as mathematical theories. At the same time, pure mathematics grew exponentially as it became ever more rarefied, a highly theoretical pursuit accessible to a select few. Abelian varieties, homological groups, linear operators, and functors were just a few of the abstruse mathematical concepts that held mathematicians' fascination during these years. High modernist mathematics, in both its pragmatic and aesthetic incarnations, split the difference by offering a common rationale to both the pragmatists and the airy theorists: the axiomatic method.

Mathematics thus exemplifies a common feature of American midcentury thought. In the world of Abstract Expressionist artists and their formalist critics, as well as in midcentury social scientists' emphasis on general theories, abstraction helped support the Cold War's "ideology of no ideology" while also maintaining a positivist stance.[76] Throughout this book, I appeal to the Cold War as a way of identifying the specific conditions of knowledge that characterized intellectual life in postwar America. Scholars have debated the degree to which the geopolitical rivalry between the United States and Communism truly defined intellectual activity in the postwar period. Modernization theory might be explained as a tool in the global struggle between the two, but can the same be said about the rise of cognitive psychology? Such questions are important and have rightly received much scholarly attention among both historians of science and historians of the social sciences.[77] My emphasis when it comes to mathematics, however, is on the new funding opportunities created in the academy and beyond by the Cold War (for both research and education) and the social and political repression that accompanied them.[78] McCarthyism, along with the need to provide a justification or rationale for the disbursal of federal funding, accounts for why abstraction became endemic to the Cold War.[79]

Focusing on abstraction and axiomatics, this book seeks to bridge the oppositional growth of postwar mathematics toward the utilitarian and the idealistic. However, my aim in doing so is broader, for the converse is also true: mathematics offers a unique perspective on the midcentury intellectual landscape. From this vantage point, high modernist humanities and sciences appear less as competing worldviews and more as two facets of the same governing epistemological commitment.

Over the course of the twentieth and twenty-first centuries, mathematical and computational rationales have become increasingly embedded in theories of knowledge, modes of governance, and new technologies. Historians have documented how statistical knowledge became integral to the functioning of the modern American state, from social security to managing the economy, measuring unemployment, counting the population, and managing risk. Others have noted the growing influence in the postwar period of mathematical techniques in economics, the biological sciences, and the social sciences. Finally, more recently, scholars have begun to scrutinize the overwhelming influence of algorithmic thinking on every aspect of our daily lives, occasioning a growing demand to reopen the "black box." In focusing on the history of axiomatics, my argument is that explaining these broad transformations requires a better account of these technical fields themselves. Too often, statistics, mathematics, and even computer science are treated as a collection of theories that are simply applied to a given context. However, each one of these technical fields has its own disciplinary matrix, philosophical premises, and active controversies. If we are to come to terms with their growing influence, then it behooves us to better understand their histories.

Chapter Outline

Chapter 1, "Pure Abstraction," centers on the world of pure mathematics. Pure mathematicians sought to build upon the theoretical inroads made in algebra in the 1920s and extend them to other areas of study, from topology and geometry to number theory. In the process, high modernist mathematics served as a force of unification and generalization. For the numerous mathematicians who joined the fold in the postwar decades, mathematics was no longer divided into logic versus intuition, as it had been in the nineteenth and early twentieth centuries. Instead, this new generation was indoctrinated into a world in which the most salient distinction among mathematicians was the one separating those who sought to construct unifying theories from those who were busy solving concrete problems.

For pure mathematicians, however, both abstraction and universality were moving goalposts. A topological theory that mathematicians celebrated as the pinnacle of the abstract approach at one moment could be replaced by yet another, more abstract and universal theory in a matter of years. Thus, following Fredric Jameson, I take high modernist mathematics to be an ongoing project of rewriting. With high modernist mathematics, such rewriting became a mathematical discipline par excellence. The chapter begins by looking at the life of an individual mathematician, Norman Steenrod. As I show, his professional development captures the reorientation of the field toward generalized and axiomatic theories. Zooming out, the chapter turns to the growth of topology, in which Steenrod played an important role, to demonstrate how mathematicians, in their enthusiasm for rewriting, repeatedly inverted the relation between structure and content. Finally, expanding further, I turn to category theory to illustrate how the same abstractionist logic functioned on a field-wide level. The impetus and pace for rewriting was so strong that by the 1960s many in the older generation of pure mathematicians turned against their younger colleagues, accusing them of no longer being able to recognize their own contributions to the field because of how radically they had been reworked.

Chapter 2, "Applied Abstraction," turns from the realm of pure mathematics to the world of applied mathematics. The chapter tracks the changing meaning of mathematization before and after World War II by closely examining the writing of Oswald Veblen and George Birkhoff. As I demonstrate, the proliferation of axiomatics together with Einstein's theory of relativity convinced Veblen and Birkhoff that the role of mathematics in explicating natural phenomena was based in theoretical construction. Veblen and Birkhoff had a profound influence on the generation of American mathematicians who found themselves involved in defense research during World War II. The proliferation of new fields during the war, from operations research to communication theory and computing, further cemented the idea that the utility of mathematics was in no way limited to measurement and calculation.

After World War II, the relation between mathematics and theory construction was best illustrated in John von Neumann and Oskar Morgenstern's *Theory of Games and Economic Behavior*. Besides introducing its readers to game theory, the book also offered a careful guide to axiomatic reasoning. In examining national reports produced by the mathematical community and the writings of several influential mathematicians, the chapter demonstrates how this new conception of mathematization became the dominant one in the postwar moment.

Chapter 3, "Human Abstraction," follows the axiomatic conception of theory as it was adopted in the postwar period by a host of social scientists. In the aftermath of World War II, the Social Science Research Council (SSRC) sponsored several initiatives to promote the use of mathematics in the social sciences. Such initiatives were not new, but the inclusion of axiomatics as a mathematical topic necessary for budding social scientists marked a shift. Axiomatics was promoted by mathematically trained individuals such as Robert M. Thrall and Howard Raiffa, who turned to operations research and game theory in the aftermath of the war. However, as I show, the assertion that mathematics did not equal quantification was shared by many social scientists during the period, even those with no formal mathematical training.

A diverse set of scholars including Claude Lévi-Strauss, Herbert Simon, Paul Lazarsfeld, Kenneth Arrow, and George A. Miller argued that the utility of mathematics in explicating social phenomena was not limited to measurement and calculation. It was *mathematical thinking*, they claimed, that was necessary, and that offered conceptual clarity.

In examining the writings of these social scientists, I illustrate one of the more lasting influences of axiomatic thinking in the postwar social sciences: the separation between the analytic or formal conception of a theory and the social phenomenon it described. Theories might have emerged from the world, but their formal descriptions existed independently. For social scientists, this break called into question the meaning of "theory" in their field. Neither predictive nor inductive, theory for them was positive and explanatory.

Chapter 4, "Creative Abstraction," returns to the world of pure mathematics to ask how mathematicians were able to maintain such a high degree of autonomy during the postwar period. Mathematicians insisted that their work belonged among both the sciences and the humanities. Mathematical activity, according to mathematicians like Marston Morse, was akin to art. This chapter offers a three-part analysis of the mathematical discourse of the time, following national reports, articles, and speeches by pure mathematicians in the postwar moment. First, I suggest that, in their writings and public outreach, mathematicians called upon art to position pure mathematics outside of the demands of the day. For them, art offered the only real corrective to the postwar utilitarian spirit. Second, mathematicians like Paul Halmos appealed to modernist art as an epistemological double. In no other field of study was the break between the world and the work so complete. I thus follow the similar strategies that artists and pure mathematicians developed for justifying their work to themselves and others. Several mathematicians insisted that the aesthetic dimension of mathematics was proportional to its separation from the world. The more abstract and free a given theory was, the greater role aesthetic considerations played in its development. These mathematicians thus adhered to a similar conception of aesthetic autonomy as espoused by Clement Greenberg, one of the postwar era's more celebrated modernist critics.

Finally, such discursive analysis of mathematicians' commitment to an artistic image of their field helps account for the ubiquity and appeal of abstraction at midcentury. As I show, mathematicians often credited axiomatics for rendering mathematics a creative and artistic pursuit. In other words, they pointed to axiomatics to account for *both* the utilitarian and the nonutilitarian aspects of modern mathematics. While this might sound like a contradiction, it was not the case for mathematicians, who believed the two were simply two sides of the same coin. Mathematics thus lays bare the contradictory (and hence productive) nature of abstractionist thought.

Chapter 5, "Unreasonable Abstraction," turns to the relation between pure and applied mathematics. The chapter follows the disputes among mathematicians that enlivened the community in the postwar period. I survey how leaders of the American Mathematical Society appealed to abstraction to counter the changing funding regime and institutional formation brought on by the Cold War. Mathematicians like Marshall Stone, Marston Morse, and Adrian Albert questioned the postwar regime of basic and applied research by demonstrating the incompatibility of these categories with mathematical research. By stretching the concept of utility, mathematicians were able to maintain remarkable autonomy over the field in the postwar period.

The chapter asks how postwar mathematics might challenge some historians' deeply held ideas about the sciences during the Cold War. As I demonstrate, the postwar abstractionist theory of knowledge had its most direct impact on graduate training in applied mathematics, which remained a source of frustration to applied mathematicians throughout the period. Despite the fact that the increase in funding for both research and education in mathematics was motivated by the application of mathematics, it was pure mathematics that dominated academic institutions at the time.

Finally, in Chapter 6, "Historical Abstraction," I examine how mathematicians turned from philosophy to history in the postwar period to answer the question *What is mathematics?* I trace a series of disputes that emerged between historians of mathematics and historically minded mathematicians in the 1970s, most notorious among them a controversy between historian Sabetai Unguru and leading mathematicians such as André Weil that resulted from the publication of Unguru's "On the Need to Rewrite the History of Greek Mathematics." As I show, these disputes can be best understood in relation to two major historiographic reorientations at the time: Thomas Kuhn's publication of *The Structure of Scientific Revolutions* in 1962 and Quentin Skinner's publication of "Meaning and Understanding in the History of Ideas" in 1969. The problems facing historians of mathematics, I argue, point to an inherent tension in any conceptual history or history of ideas between the permanent and the transitional. I ask how the history of mathematics speaks to a history of ideas that acknowledges *continuity* while rejecting Platonism.

Axiomatics centers on the world of midcentury mathematicians, but its scope extends beyond it. By focusing on abstraction in its many valences, the book queries the relation of modernism to mathematics, in particular the impact of mathematics on American modernism. While modernism is notoriously difficult to define, its entanglement with abstraction is undeniable. In art, as in science, abstraction captured an uneasiness with, and a questioning of, the relation of representation to reality, a problem that was endemic to the modernist project. The centrality of abstraction in mathematics, as both an object and a method, makes mathematics the crucial place from which to probe abstraction's rich history and its claims to modernism. During the Cold War, abstraction was symptomatic of a high modernist inclination: Philosophically, it underwrote the semantic or analytic traditions. Politically, it guaranteed intellectual and artistic autonomy. Culturally, it designated a dawning age of American ascendency. Scientifically, it questioned the essence of theoretical knowledge. Nowhere are these connections more readily apparent than in the world of midcentury American mathematicians.

CHAPTER ONE

Pure Abstraction

Mathematics as Modernism

In every branch of mathematics there is one plane of generality on which the theorems are easiest to prove, and needless complication arises as quickly by falling short of this as by exceeding it. It is a mark of the great mathematician to have taken a number of separate theories, fragmentary, intricate and tortuous, and by a profound perception of the true bearing and weight of their methods to have welded them into a single whole, clear, luminous, and simple.
—Norbert Wiener

In 1950, the Department of Mathematics at Harvard University sought to hire a new senior mathematician into its ranks. In the final analysis, two mathematicians rose to the top, French mathematician Claude Chevalley and Swedish mathematician Arne Beurling. Both Chevalley and Beurling had earned their PhDs in 1933, and by 1950 both were well-respected among the mathematical community. Chevalley, who had arrived in the United States in 1939 as a visiting scholar at the Institute for Advanced Study (IAS), was a professor at Columbia University. Beurling, who had just completed a year as a visiting professor at Harvard, was a professor at Uppsala University. Initially, the department's vote was not unanimous. Only after additional deliberation did the faculty decide to unite their support behind Chevalley.[1]

A founding member of Bourbaki, Chevalley was a celebrated algebraist whose work represented the cutting-edge approach to mathematical research at the time—the abstract approach.[2] Writing in support of Chevalley, Oscar Zariski emphasized: "Chevalley's mathematical work is characterized by many-sidedness of content, originality of abstract conception and a formalism of superb craftsmanship. A clear and precise thinker, he has, to the highest degree, the gift of discovering the abstract scheme of things that lies behind concrete and difficult mathematical problems."[3] His penchant for abstraction and his ability to see beyond accumulated detail elevated Chevalley among his peers. His numerous publications in algebra and algebraic geometry had earned him the deep respect of the wider mathematical community.

An expert in analysis, Beurling's stature among the international community did not shrink beside Chevalley. He was invited in 1950 to give one of the plenary talks at the International Congress of Mathematics, which alternately took place at MIT and Harvard. However, unlike Chevalley, Beurling represented a different approach to mathematical research, one which mathematicians referred to as "classical" in order to distinguish it from the modern emphasis on abstraction. Neither a pioneer nor an early adopter of modern algebraic techniques, Beurling's research focused not on theoretical constructions but rather on solving problems whose origins were first posed in the mathematical work of nineteenth-century mathematicians.

Harvard mathematician Lars Ahlfors was one of the three faculty members who originally voted for Beurling. In a letter to the faculty, he explained his reasoning:

> To make a comparison between Beurling and Chevalley is impossible already for the reason that they represent opposite *types of mathematicians* and opposite fields in mathematics. Chevalley is very abstract with broad knowledge and a taste for sweeping generalizations . . . Beurling is pure analyst, he takes no interest in the abstract approach and his techniques do not allow broad generalizations. His force is a penetrating analysis aimed at the unearthing of hidden connections which could not be discovered by the most powerful use of formalistic arguments.[4]

It was not just that Beurling and Chevalley published in different mathematical fields. More fundamentally, according to Ahlfors, they represented two different *kinds* of mathematicians and two different mathematical aesthetics.

Whereas intuition and logic had served as the decisive markers of mathematics at the turn of the century, by the 1940s, generality and concreteness had replaced them. Starting in the 1940s, what distinguished two kinds of mathematicians was not the relative role intuition or logic played in their investigations, but their respective dispositions toward abstraction.[5] For high modernist mathematicians, abstraction became synonymous with universal theories, *not*, as it was at the turn of the century, with the supposed break between mathematical research and physical reality. Generality, rather than logic, defined the abstract conception of pure mathematics in the postwar period. The generation of pure mathematicians who had joined the field in the 1930s were unconcerned with the foundations of mathematics. They cared less about ontology and epistemology and more about the dangers of increased specialization, which they believed could only be combated through generalized theories. Abstraction became a guiding principle, and a rallying cry for an entire generation of mathematicians.

Despite their emphasis on, and commitment to, universal and abstract theories, mathematicians understood neither abstraction nor unification as stable terms or as binary markers, but rather as processes. Rather than denoting a series of oppositions between form and content, the general and particular, the theoretical and the concrete, the abstract and the representational, mathematicians approached abstraction as an ongoing project. As will become clear in this chapter, a monograph that mathematicians celebrated as the pinnacle of the abstract approach in 1942 could and often would be supplanted in a matter of five years by another text claiming to offer yet another, more abstract and more unified presentation. It was the tireless search for a unified theory, celebrated by some and held in suspicion by others, that defined high modernist mathematics. More than an adjective, *abstract was a verb*.

This unwavering striving for abstraction and universality implied that for mathematicians, modernity itself was continuously in the making. This was nowhere better exemplified than in the publication of Bartel L. van der Waerden's *Modern Algebra*. When the textbook was published in 1930 it was celebrated as the culmination of the abstract and generalized approach to algebra, and it served as an inspiration to young mathematicians new to the field. Yet twenty-nine years later, when a new edition of the book was released, the word "modern" had been dropped from the title. Despite including revisions and additional material, the new edition was simply titled *Algebra*. This was not a unique phenomenon. Saunders Mac Lane and Garrett Birkhoff's *A Survey of Modern Algebra*, which came out in 1941, similarly elided the word "modern" from its title when it was republished in 1960. Like abstraction, mathematical modernity was an ongoing project.

As such, I see high modernist mathematics as the modernist inclination pushed to its logical conclusion. The crisis of referentiality which had troubled turn-of-the-century mathematicians, artists, and scholars had, by midcentury, transformed into a mathematical infatuation with abstraction as both object and method. In no other field was the commitment to abstraction and universalism so totalizing as it was in pure mathematics, where it restructured the field on every level. One of the four maxims of modernity, writes critical theorist Fredric Jameson, is narrative rewriting, which seeks to replace previous narrative paradigms. "Modernity is not a concept, philosophical or otherwise, but a narrative category," he writes.[6] Jameson explains that the "trope of 'modernity' is always in one way or another a rewriting, a powerful displacement of previous narrative paradigms."[7] For Jameson, any attempt to define modernity as such entails rewriting an older narrative. Moreover, according to Jameson, all the features typically invoked as markers of modernity (e.g., self-consciousness or attention to language) are merely pretexts for the actual dynamic driving modernity, which is the "registering of a paradigm shift" and its attendant moods.[8]

Jameson has in mind theorizers of modernism, but his analysis is also remarkably applicable to high modernist pure mathematics. Mathematicians might have pointed to formalism and the axiomatic approach as the defining features of modern mathematics, but their high modernist sensibility is epitomized by their perpetual drive to reconstruct old theories anew. Thus, following Jameson's observations, I take high modernist pure mathematics to be an ongoing project of rewriting. With high modernist mathematics, rewriting became a mathematical discipline par excellence. Fundamentally, this act of rewriting was not as a post hoc account of mathematicians' research, a reinterpretation of past work. Rather, it constituted mathematical research as such. It is this processual quality that makes mathematics a crucial element of the history of modernism. Not only were changes in mathematical thought indispensable to the wedding of modernism and abstraction at the turn of the century, but by the mid-1950s the modernist project was pursued most forcefully in mathematics. With each act of rewriting, the relation between figure and ground, the abstract and the concrete, the signifier and the signified came undone. As mathematician Richard Courant explained:

> One must bear in mind that the terms "concrete," "abstract," "individual" and "general" have no stable or absolute meaning in mathematics. They refer primarily

> to a frame of mind, to a state of knowledge and to the character of mathematical substance. What is already absorbed as familiar, for example, is readily taken to be concrete. The words "abstraction" and "generalization" describe not static situations or end results but dynamic processes from some concrete stratum to some "higher" one.[9]

The binaries that characterize mathematical research were always in the making.

In what follows, I enumerate three conceptions of mathematicians' modernist rewriting: rewriting as *revising*, rewriting as *restructuring*, and rewriting as *recasting*. I begin by following the early career of one mathematician, showing how he learned to revise—to "see anew"—what mathematical research was all about. As abstractions became the dominant approach, mathematicians had to reorient their gaze by focusing not on particular examples but on what united them. Zooming out from the life of an individual to a mathematical field, I examine the development of topological theory in the postwar period. Here, rewriting functioned as restructuring. In their efforts to create universal theories that would unify the widest possible mathematical phenomena, mathematicians repeatedly inverted the relation between structure and content. A theory, once established, could always be subsumed by the construction of a new, more abstract structure. Finally, by looking at category theory, I examine how mathematicians sought a new mathematical language with which to recast the disparate subfields of mathematics into one coherent whole.

By the mid-1960s, all these various abstractionist efforts had given rise to a generational break, with some of the older mathematicians believing that abstraction had become an end in itself rather than a means to a goal. Mathematical *meaning*, they argued, had been evacuated.

The Education of Norman Steenrod

Norman Steenrod graduated from high school in Dayton, Ohio, in 1925, when he was just fifteen years old. Instead of pursuing college immediately, he followed in his brother's path to become an industrial tool designer. Two years later, he enrolled at the University of Miami, and in 1929 he transferred to the University of Michigan, where he completed his undergraduate studies. It was at Michigan that Steenrod first discovered his mathematical talent and came under the wing of Raymond Wilder.

During his senior year, Steenrod started attending graduate courses and was even assigned his own research problem in topology by Wilder. In June, after he graduated, Steenrod went back to Ohio in search of a job. The year was 1932, and with the unemployment rate rising to more than 20 percent, there was not much demand for tool designers. With free time on his hands, Steenrod continued to work on his research problem. He took preprints of mathematical papers with him while job hunting, working through them while waiting to meet potential employers. Throughout the summer and the winter, he reported back to Wilder on his progress, sending him detailed letters on his successes and failures.

Wilder had received his PhD under the supervision of R. L. Moore at the University of Texas at Austin in 1923. Moore is famous for developing a unique pedagogical program in mathematics at Texas. He would begin his courses by providing the students in his class with a set of axioms and then he would let students explore and prove theorems on their own.[10] He would even remove mathematical textbooks from the campus library to ensure students were not consulting other sources. His work and that of his students focused on the area of point set (or set-theoretic) topology.[11] Unlike geometry and algebra, topology in the 1920s was a fairly young field of research dating back to the work of Henri Poincaré at the end of the nineteenth century, and originally known as analysis situs.[12]

During the 1930s there were two competing approaches to the study of topology in the United States. The first, represented by Moore's school, focused its research on concepts such as connectedness, continuity, and compactness. The second, centered at Princeton, was known as combinatorial topology (and later as algebraic topology), which used combinatorics to study invariants between topological spaces.[13] The former was based on direct geometric concepts, such as continuity, connectedness, and closeness, which were defined rigorously but were rooted in immediate geometrical intuition. For example, following an intuitive understanding of topology, to say that a curve is continuous is to say that one can draw the curve without lifting one's pen off the page. In his own work, Wilder was trying to strike a balance between the two approaches, but in his teaching he focused on the set-theoretic aspect of topology. It was the geometric approach to topology that first attracted Steenrod to mathematics and which compelled him to continue his mathematical studies.

Wilder must have been truly impressed by Steenrod. Over the following decade he maintained a close correspondence with Steenrod, supporting and advising him on matters large and small.[14] When Steenrod, with Wilder's encouragement, decided to apply to schools for graduate education, Wilder sent letters on his behalf, extolling his "creative ability," "capacity for research," "delightful personality," and "cheerful disposition."[15] And it worked. By April 1933, Steenrod had received fellowship offers from both Harvard and Princeton. Uncertain at first, he eventually decided to attend Harvard. In June 1933, just before he moved to Cambridge, Massachusetts, Steenrod wrote Wilder from Ohio informing him that he was currently considering the problem "When is a continuous curve the sum of a finite number of acyclical curves?"[16] He sent Wilder some of his early findings on the topic, as well as a list of questions that were "bothering" him and to which he hoped Wilder might have an answer. Considering his later work, for which he became well-known in mathematical circles, what is surprising is the amount of careful illustration that accompanied the text (see Fig. 1.1).

In August 1933, Steenrod finally managed to secure a job as a die designer at the Chevrolet plant in Flint, Michigan. Frustratingly, he managed to work for only three weeks before having to move, and managed to save only $60 for the coming year at Harvard. Once he arrived at Harvard, Steenrod's interests began to change. In his first semester, Steenrod attended a seminar on analysis situs taught by mathematician Marston Morse, who at the time was still teaching at Harvard but would soon take up a new position at the IAS. Topology was one of the mathematical fields Steenrod was most familiar with, and yet he quickly learned that he had a lot to learn. "[Morse] has no strong interest in curve theory," he wrote to Wilder, "his interest is mainly in combinatorial theory."[17] Combinatorial topology was more complex and less intuitive than set-theoretic topology, and as Steenrod quickly found out, the learning curve was steep.

During his first year at Harvard, Steenrod continued pursuing his interest in set-theoretic topology. Instead of sharing his work with his professors, he sent it to Wilder back in Michigan. "I feel somewhat peculiar about sending you this stuff," he wrote. "But, you see, it is this way: I feel that I have proved something, and have a hankering to tell someone about it; no one here at Harvard seems the least bit interested, so what can I do but send it to you?"[18] When he did try to present his work to Morse, his efforts were rebuffed. "Morse (becoming annoyed with my slithy-shady methods) has set me the task of crawling through Veblen inch by inch."[19] Published in 1921, Oswald Veblen's *Analysis Situs* was the standard introductory text in English to the basic concepts in combinatorial topology. For young mathematicians at the time, the book was a necessary entry point one had to pass through in order to contribute to the field.[20]

2

The fact that continua containing no continuum of condensation turn out to be the sum of two acyclic curves suggests that this might also be true of all curves of a more general class. This is not so; for the nearest classification of curves above continua lacking continua of condensation is the class of regular curves, and the following is a regular curve which fails to be the sum of even a countable number of acyclic curves:

Ex. 1

Omit interiors of successive triangles

It is evident that the im-kleinen cycle points are everywhere dense in M, consequently the first derived aggregate of M with respect to acyclic curves is M itself, and M is equal to all its derived aggregates. And, since C_α^β is not zero for any ordinal β of the first or second class, M is not the sum of any countable number of acyclic curves.

Next I thought that the sum of a finite number of acyclic curves might be an already classified type of curve. At first I thought it must always be a perfect continuous curve; but the following is an example of a continuous curve not perfectly continuous which is the sum of two acyclic curves:

Ex. 2

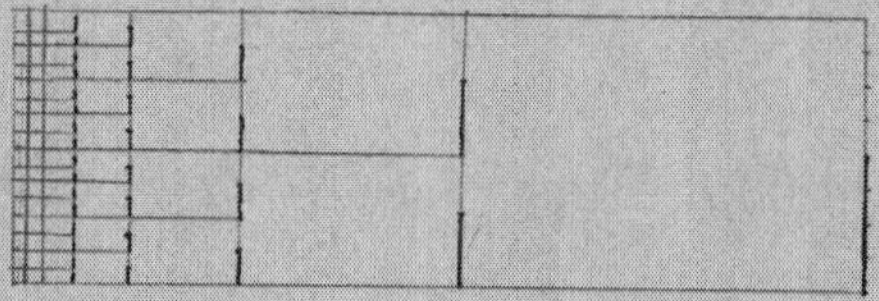

The acyclic curve A_1 consists of all the horizontal lines plus the vertical base line and all vertical inked lines. The acyclic curve A_2 consists of all horizontal lines plus the vertical base line and all vertical penciled lines. The above curve is rational. I've been wondering if a finite acyclic curve-sum is always rational. Or, more generally, is the sum of two rational curves always rational? I've spent a good bit of effort over this point, and I shall be irritated with myself if it is easily proved. I have the idea that a curve M which fails to be rational at some point must contain a subcontinuum K which has no local separating point; but, if this is so, K will appear in every derived aggregate of M, and M will fail to be the sum of any countable number of acyclic curves.

FIGURE I.1. Two pages form Steenrod's letters to Wilder, which include his handmade drawings. Such drawings are completely absent from Steenrod's later work.

Source: Raymond Wilder Papers, 1914–1982, Dolph Briscoe Center for American History, University of Texas at Austin

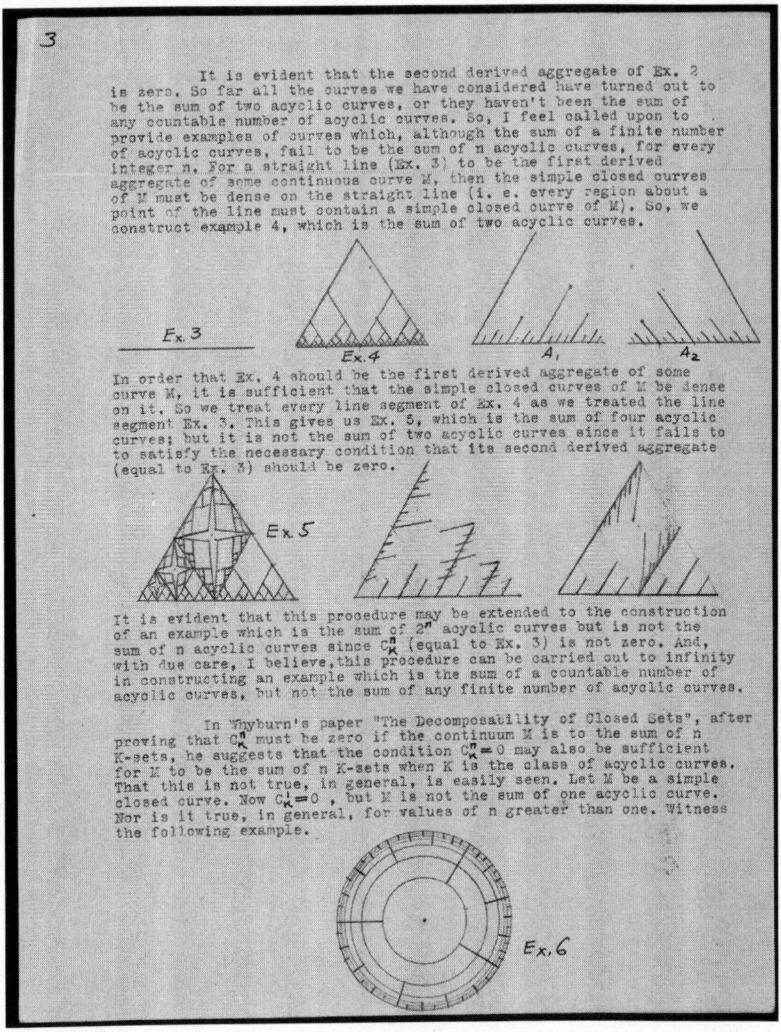

3

It is evident that the second derived aggregate of Ex. 2 is zero. So far all the curves we have considered have turned out to be the sum of two acyclic curves, or they haven't been the sum of any countable number of acyclic curves. So, I feel called upon to provide examples of curves which, although the sum of a finite number of acyclic curves, fail to be the sum of n acyclic curves, for every integer n. For a straight line (Ex. 3) to be the first derived aggregate of some continuous curve M, then the simple closed curves of M must be dense on the straight line (i. e. every region about a point of the line must contain a simple closed curve of M). So, we construct example 4, which is the sum of two acyclic curves.

Ex. 3 Ex. 4 A_1 A_2

In order that Ex. 4 should be the first derived aggregate of some curve M, it is sufficient that the simple closed curves of M be dense on it. So we treat every line segment of Ex. 4 as we treated the line segment Ex. 3. This gives us Ex. 5, which is the sum of four acyclic curves; but it is not the sum of two acyclic curves since it fails to to satisfy the necessary condition that its second derived aggregate (equal to Ex. 3) should be zero.

Ex. 5

It is evident that this procedure may be extended to the construction of an example which is the sum of 2^n acyclic curves but is not the sum of n acyclic curves since C^n_K (equal to Ex. 3) is not zero. And, with due care, I believe, this procedure can be carried out to infinity in constructing an example which is the sum of a countable number of acyclic curves, but not the sum of any finite number of acyclic curves.

In Whyburn's paper "The Decomposability of Closed Sets", after proving that C^n_K must be zero if the continuum M is to the sum of n K-sets, he suggests that the condition $C^n_K = 0$ may also be sufficient for M to be the sum of n K-sets when K is the class of acyclic curves. That this is not true, in general, is easily seen. Let M be a simple closed curve. Now $C^1_K = 0$, but M is not the sum of one acyclic curve. Nor is it true, in general, for values of n greater than one. Witness the following example.

Ex. 6

FIGURE 1.1. *(Continued)*

The pressure to learn and adopt the combinatorial method was strong. Not only Morse, but topologist Hassler Whitney, who had obtained his PhD from Harvard only a year earlier, also impressed on the young Steenrod the importance of the algebraic method and encouraged him to supplement his study of topology with group theory. After he finished reading

Veblen's book on his own, Steenrod still felt lost in the material. "Relative to the combinatorial analysis situs," he wrote in February, "something of a fog enshrouds me whenever I think about it. I neither know what I know, nor what I don't know."[21] Thus, together with four other students, he set about to read the book once again.

Steenrod was none too happy at Harvard. He found it difficult to get the professors' attention and made few friends. "Social life, here, is nil," he wrote. "I could do better work with little more of it. And the town of Cambridge is very much an eye-sore."[22] In the spring, when offers came pouring in from other schools, Steenrod decided to leave Harvard to continue his studies at Princeton, which at the time was the mathematical capital of the United States. Yet his year at Harvard had certainly been enlightening. With little else to do, Steenrod had fully devoted himself to mathematics. In June, just before he traveled back to Ohio to find a job for the summer, he reflected upon his accomplishments: "This year has been plenty successful for me. I've had my brain turned upside down, shaken, and put in order again. Something of a mental renovation. I believe I now know a bit more about mathematics."[23]

The "mental renovation" involved, as Steenrod acknowledged toward the end of his letter, a complete reorientation of his research approach: "I guess I've been temporarily weaned away from my beloved curve theory."[24] Steenrod added that he planned to return to his prior research during the summer. A year later, by the time he had finished his first two semesters at Princeton, his interest in curve theory had completely fallen by the wayside. The transformation that he had begun at Harvard only accelerated at Princeton, which was the center of algebraic topology at the time.

Princeton in the 1930s stood in stark contrast to Harvard. Unlike Harvard, socializing was part of the course of study. The common room had been outfitted as a game room where students and faculty alike spent much of their time playing games (the most popular ones were *Kriegspiel* and Go). Students were able to interact not only with Princeton faculty but also with those scholars affiliated with the newly established IAS, and the list of renowned mathematicians in the greater Princeton area was unmatched. As he was preparing to relocate, Wilder sent Steenrod a letter warning him not to get too tempted by the mathematical bounty with which he would soon be confronted: "Let me advise you *do not try to attend too many lectures*. You will find it very difficult to resist the temptation—there will be loads of seminars starting up, and you'll want to attend them all. But it's no use, it can't be done. Possible you may shop around a little at first, but try to settle down to a normal load after a while."[25] Steenrod not only had the chance to study under some of the leading mathematicians of the time and keep track of the most up-to-date developments in the field, but he also came to know many of the mathematicians who in the 1940s and 1950s would become the elite of the American mathematical community.[26]

At Princeton, Steenrod quickly fell under the influence of Solomon Lefschetz. He was among the most prestigious American mathematicians working on algebraic topology and was, in fact, the first to introduce the word "topology" into English, in his 1930 book of the same name. Lefschetz had an outsized personality. In mathematicians' recollections of the time, stories about Lefschetz abound. When he attended a seminar, no mathematician young or old was immune to his criticism. He would routinely interrupt a speaker to ask questions or argue about a given point in a proof, a fact that Princeton's students memorialized in a little rhyme: "Here's to Lefschetz, Solomon L./Irrepressible as hell/When he's at last beneath the sod/He'll then begin to heckle God."[27]

A few months into his first semester at Princeton, Steenrod began working on a new topological problem. The idea came to him after he stumbled upon a counterexample to an open problem in topology. The case intrigued him, and he began to wonder if he could find a similar counterexample in higher dimensions. Seeking advice, he approached Lefschetz with this problem. "Well, Lefschetz launched into a lecture," he reported to Wilder. "He discussed the matter of young mathematicians acquiring the habit of publishing numerous papers on trivial problems. It appears that the true Princetonian method is to work only on *general* problems and to publish only when some step in theory has been accomplished."[28] Lefschetz went on to ridicule other mathematical schools such as the Polish topological school and Moore's school at the University of Texas.[29] For all their work, he told Steenrod, they had nothing to show. The message was clear: "I couldn't possibly hope to become a great mathematician by publishing a lot of papers on trivial problems."[30] Namely, it was not solving concrete cases but, instead, making inroads in the direction of theory construction that differentiated the great mathematician from lesser ones.

Steenrod was nonplussed by Lefschetz. "It was a good lecture," he reported, "and I agreed that his main thesis was true."[31] Indeed, considering Steenrod's subsequent work in topology, he clearly took Lefschetz's advice to heart. And Steenrod was not the only one: years later, Edward McShane, who was a postdoctoral fellow at Princeton in the 1930s, recalled that Lefschetz used to refer to Moore's topological school as the "Concerning School." The reason was that many of the papers published by Moore and his students bore the title "Concerning X," where X was a particular case the author examined. For example, from 1919 until 1945, fourteen out of the fifteen papers Moore published were of this type (e.g., "Concerning Simple Continuous Curves"). According to McShane, Lefschetz used to say that "to write a book about topology and confine oneself to this subject matter, is like writing a book about zoology and confining oneself to rhinoceros."[32] What Lefschetz questioned was the proper level of mathematical analysis and what counted as progress. Neither topology nor zoology, he maintained, could be reduced to the sum of its component parts. Zoology was not the study of rhinoceroses, elephants, giraffes, and lions, and topology was not just a collection of individual results regarding curves and surfaces. To count as true theoretical knowledge, a result had to speak to several individual cases. What Lefschetz sought to define was the meaning of theory in mathematics.

Other scientists and social scientists similarly questioned the meaning and limits of theoretical knowledge, but what distinguished pure mathematics was that it did not have an experimental component. The distinction thus was not between theory on the one hand and experimentation on the other, but between problem solving and theoretical construction.[33] With Lefschetz, theory was wed to generality, a process which only increased in the following decades. Lefschetz supervised sixteen students during the 1930s and 1940s, many of whom played fundamental roles in the development of mathematics in the following decades by setting up new research fields and supporting the institutional development of postwar mathematics.[34] Beyond his students, his influence on the mathematical community at Princeton cannot be overemphasized. Outside of Princeton, Lefschetz served as an editor of the *Annals of Mathematics* for twenty-five years, where he continued to impress his standards and transform the journal into a top-tier mathematical publication.

Mathematicians relied on the remarkable transformation in algebra to achieve their generalization.[35] Upon arriving at Princeton, Steenrod wrote to Wilder with fresh bewilderment, "This place is 'group' crazy. One comes here fresh and innocent. But it is not long before you go around babbling about groups."[36] Steenrod was surprised by the ubiquity of group theory in almost every class and seminar he attended. Hermann Weyl was giving a course on continuous groups, as was Luther Eisenhart, albeit from a different perspective. James Alexander was lecturing on abelian groups, and Leo Zippin provided an analysis of torsion groups. If that was not enough, Steenrod teamed up with two other students for a personal seminar on group theory. He wanted to ensure he did not "grow weak on group theory."[37] The popularity of group theory in the 1930s marks the beginning of high modernist mathematics. Mathematicians first used the term "group" within very specific contexts either in geometry or algebra.[38] For example, when examining the symmetry of a given geometric object, they asked, What are all possible transformations of a square (e.g., flipping over the diagonal, rotating by ninety degrees) such that the square remains the same? Over time, however, they recognized that a certain degree of regularity exists across the various cases and that they could define a group abstractly independent of any specific instantiation. For a while, they continued to use the concept of a group only when discussing particular problems, but by 1930, groups themselves, absent any specific context, became an object of interest for mathematicians.

The change was partially enabled by the adaptation of the axiomatic methods to the study of algebra. Crucial to this transformation was the work of German-born mathematician Emmy Noether. The daughter of a mathematician, Noether distinguished herself early on by her mathematical abilities. In 1915, she was personally invited by David Hilbert and Felix Klein to the University of Göttingen. Despite her impressive credentials, the university did not approve her appointment. As a woman, her courses were listed under Hilbert's name, and she was not officially appointed as a Privatdozent until 1919. As Leo Corry explains, Noether's work was distinguished from that of her predecessors by a conceptual change in emphasis: "Noether's abstractly conceived concepts provided a natural framework in which conceptual priority may be given to the axiomatic definitions over the numerical systems considered as concrete mathematical entities."[39] Noether considered algebraic concepts from the most general point of view. If an algebraic theory had once maintained its meaning through its referential connection to arithmetic, Noether's work dispensed with this requirement. Instead, she gave greater importance to the axiomatic definitions of these concepts. In the process, she transformed algebraic concepts into full-fledged objects and the chief subjects of algebraic theory.

A year before Steenrod arrived at Princeton, Noether, who was Jewish, immigrated to the United States. With help from the Rockefeller Institute, Bryn Mawr College gathered sufficient funds to appoint Noether as a research professor. In August 1933, Anna Pell Wheeler, a mathematician at the college, urged Veblen to contact Noether and give "Bryn Mawr as good a recommendation as possible."[40] After Noether settled in Pennsylvania, she was invited to lecture at the IAS. She arrived at Princeton for each lecture wearing the same shiny blue outfit and was notorious among the young mathematicians there for using a wet sponge to erase her blackboards.[41] In the male-dominated culture of Princeton at the time, she was referred to as "Herr Noether," not as a sign of disrespect, but of admiration. Noether's ideas were popularized in 1930 with the publication of Van der Waerden's *Modern Algebra.*

A student of Noether, Van der Waerden presented algebraic theory from a unified perspective. Reflecting on the impact of the book years later, Garrett Birkhoff explained, "It is not too much to say that the freshness and enthusiasm of his exposition electrified the mathematical world—especially mathematicians under thirty like myself."[42] Steenrod was one of those young mathematicians. In the 1930s, he and his colleagues realized that regardless of their own area of research, a deep knowledge of algebraic theory was required, as algebraic notions quickly infiltrated other fields such as topology and geometry. By the 1940s, the dominant approach to topology was thoroughly algebraic. Indeed, Steenrod's own research would be influential in promoting this transformation and transferring it to the next generation.

It took Steenrod only three years to complete his PhD, and by the time he graduated his conversion was complete. He had learned to see anew. Gone was his "beloved" curve theory.[43] His approach was now completely algebraic. Steenrod remained at Princeton for three more years as Lefschetz's assistant. In 1939, he took a position at the University of Chicago. Three years later, with Wilder's encouragement and support, he moved back to the University of Michigan, where his interest in mathematics had first been ignited. Other mathematicians of Steenrod's generation experienced a similar transformation. "Analysis in the traditional sense was no longer the central concern of the most active younger American mathematicians of the 1930's," recalled Garrett Birkhoff, who entered the field at the time.[44] Mathematicians of this generation were interested in logic, abstract algebra, topology, and functional analysis and were excited to extend these new theories further. "Above all," the young Birkhoff remembered, "we hoped to be free to pursue our research into fundamental questions of pure mathematics suggested by these new ideas and techniques, and to become famous by discovering still other basic new ideas and techniques, undisturbed by the political storms which had already driven so many European colleagues to our shores in search of asylum."[45] Birkhoff makes clear that this optimism came to a quick end with the start of World War II, when many of these young mathematicians found themselves working directly on defense-related research.

For some mathematicians, the war was a turning point. They left behind their pure dreams and turned their energies to applied research. However, those who returned to pure mathematical research in the aftermath of the war continued to strive for generality and abstraction. They were ready to write a new chapter in mathematical modernity.

Restructuring Algebraic Topology

In 1942, Lefschetz published *Algebraic Topology*. At first he had intended for the book to serve as a second edition of his 1930 book, *Topology*, but the subject had changed so tremendously over the previous decade that Lefschetz inadvertently wrote an entirely new book. Upon its publication, *Algebraic Topology* was the most comprehensive account of the field to date.[46] When George Birkhoff (Garrett's father) reviewed the book in *Science*, he announced that the book represented "a culmination of the abstract phase" in the development of algebraic topology.[47] Birkhoff was so confident in his analysis that he took the opportunity of the review to comment more broadly on the "abstract character of much of contemporary mathematics."[48] The book was published in the midst of World War II, when members of the scientific community often criticized pure mathematicians for being too detached and abstruse to be of any help to the war effort.[49] This explains why Birkhoff spent part of his review defending the abstract nature of contemporary mathematics. He acknowledged that some members of the scientific public found the current tendency in mathematics toward exceeding generality suspect, but, he maintained, abstractions were a fundamental part of any scientific work. Any concept, even one that felt quite simple and concrete, such as an integer, was arrived at first through abstraction. The real test of topologists' work, as such, would be the usefulness of their theories. The significant question was not whether the theory Lefschetz presented was abstract, but "whether topological concepts [were] going to prove widely useful."[50] During the postwar period, many more mathematicians would repeat Birkhoff's insistence that abstraction was a means to a goal, but they would no longer dare to suggest that any theory represented a "culmination of the abstract phase." For high modernist mathematicians, there was always another abstraction waiting to be pursued.

Topologists, then and now, are interested in the characteristic features of a space that are invariant under continuous deformation. One way they are able to study these features is by attaching an algebraic object to a given space. As long as two spaces can be continuously deformed into one another they will have the same algebraic object attached to them (up to isomorphism). Conversely, if two spaces (e.g., a sphere and a doughnut) have different algebraic objects attached to them, they necessarily are not equivalent.[51] An analogy with Cartesian analytic geometry is useful yet misleading. Since Descartes, mathematicians have learned that they can describe a given geometrical curve or surface algebraically. Moreover, they soon realized that certain problems that seem intractable from a geometrical point of view are nonetheless easily solved algebraically. The analogy breaks down, however, since analytic geometry, unlike algebraic topology, is a two-way relation. Given an algebraic equation, I can draw the curve it describes, and given a curve, I can deduce its algebraic equation. The same is not true in the case of algebraic topology. Given an algebraic characterization of a topological space, it is impossible to deduce which space it describes. Stated somewhat differently, while the algebraic characterization of a space arises out of a space's underlying geometrical structure, once it is described in algebraic language, some geometrical information is *lost* and cannot be recovered. Moreover, whereas in Cartesian geometry the algebra involved is mostly that of polynomial equations, in algebraic topology the algebra is the abstract algebra of groups and fields.

The lion's share of topological research in the 1920s and 1930s revolved around homological algebra, the procedure by which mathematicians translated geometrical information into algebra. To do so, mathematicians appealed to the notion of complexes. In a very intuitive way, one can think of these complexes as dividing a space into smaller triangles. For example, think of taking an orange and drawing triangles on the peel until it is completely covered by such triangulation (see Fig. 1.2). By describing topological spaces in this way, topologists were able to apply algebraic reasoning to the "shape" of the surface. There was no central methodology at the time. A given procedure would be applicable to a given class of spaces, but not another. Part of mathematicians' efforts during these decades was the expansion of the number of topological spaces to which homological algebra could be applied. The development of homological algebra and algebraic topology in the postwar period offers one of the clearest demonstrations of high modernist mathematicians' commitment to rewriting.

In his book, Lefschetz strove to present homology theory from a unified perspective. He did not go into great detail about the motivation behind each theory, but rather aimed to describe each holistically.[52] Reviewing the book, mathematician Hassler Whitney noted that while the work was "written with full accuracy and with clarity . . . it [had] also largely lost touch with geometry."[53] Moreover, he added, the "elementary student" might find the book easy to read, but he or she would "get less feeling of *what the subject was all about*. Most striking of all, there [was] not a single figure."[54] In an effort to present a unified theory of algebraic topology that brought into accord the numerous developments of the previous years, Lefschetz focused on the abstract conception of the field, neglecting its geometric roots. This amounted to, following Lefschetz's own analogy, writing a book about zoology that only discussed biological theory in general without ever mentioning rhinoceroses, zebras, or any other animal.

This is why George Birkhoff believed that "most mathematicians who know something of the recent work" would agree that "topology [had] attained its approximately definitive abstract form."[55] Birkhoff explained, "The pure abstractionists have performed beautifully the essential task of giving topological ideas their appropriate abstract setting, and this has been work of the first order of importance."[56] Birkhoff, who was fifty-eight at the time, was one of the most celebrated American mathematicians of his generation. He was well versed in modern abstract mathematics, but his interests (as will be described in chapter 2) moved freely between mathematics and physics. He thus believed that now that the abstract phase of topology had reached its apotheosis, "a less abstract" phase emphasizing dynamical ideas would begin. Whereas until this point, "little [had] been accomplished by the topologists that [was] directly serviceable for application in the dynamical field," the future, he held, would prove otherwise.[57] Birkhoff passed away two years later and did not get to witness the future development of algebraic topology. If he had, he would have soon learned that, far from being over, the abstract phase of algebraic topology had only just begun, with Steenrod leading the charge.

In writing his book, Lefschetz received much help from his younger colleagues, Steenrod among them. Yet only three years after it was published, Steenrod coauthored a paper that—once again—restructured algebraic topology. When he arrived at Michigan, one of Steenrod's new colleagues was Samuel Eilenberg, another young topologist whose professional development had been cultivated by Wilder. Born and educated in Poland, Eilenberg received his PhD the same year as Steenrod, and soon after moved to

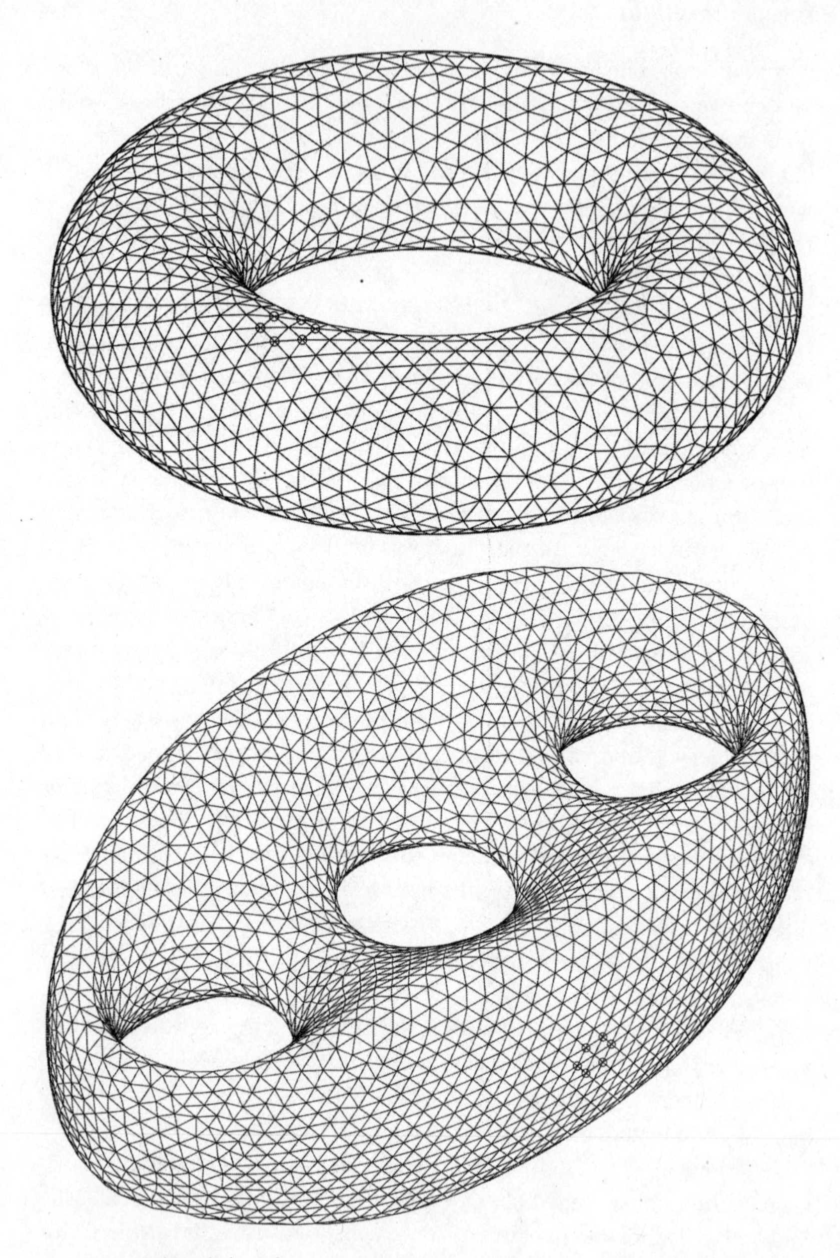

FIGURE 1.2. Triangulation of two topological surfaces. On the left is a genus 3 surface and on the right a genus 1 surface. The two surfaces are not topologically equivalent as they cannot be continuously deformed into one another.

Source: Ag2gaeh on Wikimedia Commons

Lwów, Poland, which fostered a lively center for mathematical research at the time. In 1938, with his father's encouragement, Eilenberg, who was Jewish, began searching for a way to emigrate to the United States.

It was Wilder, at Lefschetz's urging, who arranged for Eilenberg to emigrate to the United States. Although he already had three dozen papers to his name before he left Poland, the only position Wilder was able to arrange for Eilenberg was a postdoctoral fellowship.[58] A year later he was hired as an instructor in the Department of Mathematics, from which he was promoted to the rank of assistant professor in 1941. Eilenberg adjusted quickly to life in the Midwest. He played golf for the first time in his life and began attending the university's football games. "I just returned from the football game. I am frozen stiff. WE WON. This is getting monotonous," he reported to Wilder in October 1941.[59] In the apartment he shared with a fellow mathematician, he kept the icebox stocked with beer and would offer gin and whiskey to friends and visitors.

As war mobilization advanced, both Steenrod and Eilenberg enrolled in defense projects for the navy and the air force. However, their military research did not prevent them from continuing to pursue their own independent research in algebraic topology. In 1945, the two published a joint paper titled "Axiomatic Approach to Homology Theory."[60] In effect, it announced a new stage in the development of algebraic topology and practically redefined the way teachers introduced their students to the subject in the following decades. In a highly unorthodox move for a paper announcing new research, Eilenberg and Steenrod began their 1945 paper by directly appealing to pedagogical concerns. They invoked the "student of the subject," who, they explained, had to absorb a lot of preliminary details in order to begin working in the area. "The usual approach to homology theory is by way of the somewhat complicated idea of a complex," Eilenberg and Steenrod began. "Many of the ideas used in the constructions, such as orientation chain and algebraic boundary, seem artificial. The motivation for their use appears only in retrospect."[61] To avoid this "unwieldy" construction, Eilenberg and Steenrod suggested that it might be better to approach homology theory axiomatically from the top down. "It is expected," they wrote, "that an axiomatic approach or definition by properties should result in greater logical simplicity and in a broadened point of view."[62] Instead of introducing students to homology theory by effectively explaining to them how topologists developed the theory's main ideas over the years, starting with geometrical concepts before moving on to algebraic ones, Eilenberg and Steenrod suggested instead that students should first study the algebraic theory, and only then learn how it emerged out of specific topological concerns.

In 1952, Eilenberg and Steenrod put their pedagogical theory into practice in their book *Foundations of Algebraic Topology*. Here again, they accounted for their endeavor in pedagogical terms. In the preface, they offered a useful schema outlining the main steps in the *historical* development of algebraic topology:

1. Space → complex
2. Complex → oriented complex
3. Oriented complex → groups of chains
4. Groups of chains → homology groups[63]

The schema demonstrates how topologists who began with geometrical questions about the properties of topological spaces ended up with what was in practice an algebraic theory. In other words, they broke down both the historical and the methodological procedures by which mathematicians arrived at the current configuration of homology theory. Yet their goal in outlining these various steps was not to rehearse the structure of the book—on the contrary, it outlines the path not taken. Repeating the claims made in their earlier paper, Eilenberg and Steenrod explained that since students found the development of the first two steps confusing, presenting homology theory from an axiomatic perspective would sidestep these hurdles. The book, in other words, began at the end.

Since Thomas Kuhn drew attention to the role of textbooks in the formation of scientific knowledge, historians of science have turned to such books to better understand training practices, the formation of scientific disciplines, and the social and institutional formation of knowledge.[64] More broadly, pedagogy is a rich arena from which to limn how institutional, material, social, and epistemological changes intercalate cross-generationally. Steenrod and Eilenberg's appeal to pedagogy is indicative of the way in which abstraction and theory construction operated on multiple levels at midcentury. The authors appropriated a research incentive—namely, a search for unified presentation—and turned it into a pedagogical imperative. Lefschetz removed the geometrical basis from his description, but Eilenberg and Steenrod did more than this.[65] By presenting the axioms of homology first, they practically flipped the development of the theory on its head, first presenting its formal properties, and only afterward accounting for *how* and *why* the theory was developed in the first place. This point is crucial, as their approach would become a defining marker of many of the mathematical books written in the three decades following World War II. It had profound implications for the incoming generation of mathematicians who would study from those textbooks during those years.

The title of Eilenberg and Steenrod's book is clearly indebted to the title of Hilbert's 1899 landmark, *Foundations of Geometry*. Hilbert argued that the strength of the axiomatic method was its ability to clarify and make precise the assumptions and propositions that grounded a given theory. Eilenberg and Steenrod explained their motivations similarly, noting "axiomatic treatment appeared and cleared the air."[66] However, the fact that Eilenberg and Steenrod advocated their approach as a pedagogical tool demonstrates the fundamental difference between modernist and high modernist mathematics. Hilbert's *Foundations of Geometry* was never intended to be an introductory textbook to geometry. It was Hilbert's intervention into the ongoing debate surrounding the foundations of mathematics, not a programmatic statement on mathematical training. Eilenberg and Steenrod, on the other hand, had the student in mind. As such, Steenrod and Eilenberg's appeal to axiomatics represented a broader transformation among mathematicians from understanding axiomatics as a philosophical tool to promoting axiomatics as a research tool.

In presenting the axioms of algebraic topology first, Steenrod and Eilenberg were in a sense arguing that formal logical presentation provided greater clarity for students. The axioms of algebraic topology secured the foundations of the field, but they also offered students a better path into the material. At stake was not simply a question of *re*presentation but also of conceptual emphasis. Eilenberg and Steenrod did not simply offer a logical formalist reading of algebraic topology, they restructured what algebraic topology fundamentally *was*. Instead of introducing readers to concepts such as complexes, oriented complexes, and chains, which were rooted in geometrical ideas, the authors presented the axioms first and then derived these concepts from them. In doing so, they reversed conceptual priorities. Moreover, since they did not build the theory from the ground up by using examples, concrete cases merely served as illustrations of the overarching theory, which effectively made the abstract theory the core. As such, they changed what counted as the fundamental subject matter under study. They subsumed the original content and motivations under a general framework, which consequently became the subject matter of algebraic topology. This act of conceptual reversal drove mathematical research in the postwar period, and it would soon give rise to a crisis in meaning.

In 1956, four years after his book with Steenrod was published, Eilenberg coauthored another book. This time, his coauthor was French mathematician and Bourbaki founding member Henri Cartan. *Homological Algebra* covered the algebraic theory of homology independent of its original formulation in topology.[67] Over time, mathematicians came to approach homology theory as a purely algebraic theory, and the book was the first to present it as such. In the preface, the authors explained, "During the last decade the methods of algebraic topology have invaded extensively the domain of pure algebra, and initiated a number of internal revolutions. The purpose of this book is to present a unified account of these developments and to lay the foundations of a full-fledged theory."[68] By abstracting away from the original context of homology theory, the authors presented yet another self-contained theory whose applicability extended into new domains. "The unification possesses all the usual advantages," the authors remarked. "One proof replaces three," they wrote, and the theory "enjoys a broader sweep. It applies to situations not covered by the specializations."[69]

When the book came out, mathematician Saunders Mac Lane wrote a long review praising the authors for their accomplishments and summarizing some of the main results and techniques presented in the book. By way of conclusion, he made some general observations on the book's methodology: "The authors' approach in this book can best be described in philosophical terms and as monistic: everything is unified."[70] Mac Lane added that "historically, each monistic doctrine is resolved by a subsequent pluralism," and that the case at hand was no different.[71] By the time the authors finished writing their book, new research had already been published providing new cases that did not fall under their general framework. "Perhaps," Mac Lane concluded, "Mathematics now moves so fast—and in part because of vigorous unifying contributions such as that of this book—that no unification of Mathematics can be up to date."[72] This last claim defines most perfectly the high modernist project—the feeling that as soon as a theory was written down, it was primed to be rewritten by yet another more abstract and unified approach.

Recasting Universality

Lefschetz had first equated unity with theory construction as opposed to problem solving. The ability to provide an overall framework that united all the different individual studies marked the true value of mathematical research, and as Steenrod's professional training shows, beginning in the 1930s this change in emphasis oriented the career choices and research agendas of mathematicians. On an even broader level, by the end of World War II, unity applied to mathematics as a whole. It was no longer enough to construct a topological theory that unified what before had been a diverse set of studies and specialties. What was required was a governing approach that would unite all the fields of mathematics. Some mathematicians turned to rewriting the language of mathematics.

Mathematicians promoted unification to remedy what they conceived of as the increased specialization that had begun to take place at the end of the nineteenth century. They worried that as mathematics grew in size and scope, it would begin to disintegrate into an unwieldy collection of research problems and techniques, with no single mathematician able to survey the entire mathematical landscape. Mathematicians, of course, were not alone in promoting the unification of knowledge as a response to increasing specialization. Perhaps the most famous example is the Unity of Science movement in both its interwar Viennese and postwar American incarnations.[73] As Peter Galison has demonstrated, the emphasis on semantic and logical unity that characterized the early development of the movement was replaced in the postwar period with yet another conception of unity that reflected more closely the concerns of postwar scientists. The sciences of World War II, as Galison has shown, brought to the fore a unified image of science understood in terms of interdisciplinary collaborations: "The unification these [postwar] scientists had in mind was a unification through localized sets of common concepts, not through a global metaphysical reductionism."[74] The postwar unification, in other words, was fought on the ground by scientists from different disciplines collaborating with one another on specific projects.

Mathematicians understood unity *not* in terms of cross-disciplinary collaboration, but as an overarching theory of knowledge.[75] It served as a decisive impetus behind their development of new theories and a standard with which they measured new work. The clearest example of mathematicians' universalizing tendencies was the development of category theory, which in its early days was colloquially referred to as "general abstract nonsense."[76]

The theory was developed during the war by Eilenberg in collaboration with mathematician Saunders Mac Lane. The two first met at Michigan when Mac Lane delivered a series of lectures to the Department of Mathematics. Having missed the last talk, Eilenberg met personally with

Mac Lane, and the two soon realized they shared a set of interests.[77] They published their first joint paper in 1941 and followed up with eleven more by the end of the decade. When Mac Lane took over the chairmanship of the applied mathematics group at Columbia University during World War II, he recruited Eilenberg to join him. Mac Lane recalled that after they finished their war-related research for the day, the two would continue their investigations into category theory: "At Columbia, Sammy and I frequently spent the day working on Airborne Fire Control and then, in the evening, moved to his apartment to consider ordinary mathematical problems."[78] Eilenberg's confidence in axiomatization was overarching. Besides being a well-respected mathematician, he was also a well-known art dealer specializing in Indian art. After acquiring a reputation for detecting fakes, he decided to axiomatize the process. According to a colleague, "He even had a provisional list of axioms, and it was truly an elegant list."[79] Eilenberg abandoned his quest, however, after he learned that he, too, had purchased a fake. It was this sense of rigor and order that he brought to his work with Mac Lane.

Category theory has often been described as "metamathematics" because it provides a language with which to investigate various mathematical situations and recast them from one unified perspective. Thinking back to group theory is helpful here. The field grew robust when mathematicians realized they could consider certain questions regarding groups in the abstract without taking into account the particular situations from which they emerged. Besides groups, other abstract mathematical objects such as rings, ideals, and fields similarly became the general concepts of algebra mathematicians studied. With category theory, this order was reversed. Now, groups, fields, and rings became the concrete objects to which category theory was applied. The basic concepts of category theory were those of "categories" and "functors," the former denoting abstract classes of objects (e.g., normal groups) and the latter mappings between these classes.

When Mac Lane and Eilenberg sent one of their early papers to the *Transactions of the American Mathematical Society*, they worried that it would be rejected by the editor for being "too 'far out,' not really mathematics."[80] Eilenberg made sure to contact one of the editors with whom he was personally familiar and convince him to appoint as a referee a young mathematician working at the applied mathematics group at Columbia. With a little friendly influence, the paper was published. Eilenberg and Steenrod were the first to apply the notions of "categories" and "functors" to algebraic topology. They also adopted one of the primary tools of category theory, known as "diagram chasing," in their book, making it a mainstay of algebraic topology.

Over the next two decades, category theory developed into its own area of research, but its greatest influence on the growth of mathematics in the postwar period was in how it impacted already existing subjects.[81] Category theory recast mathematical theories under a unified set of concepts (categories, functors) and approaches (diagram chasing). As Leo Corry explains, "The subject-matter of category theory is the various mathematical disciplines, abstractly considered."[82] Beyond algebraic topology, category theory applied to geometry, logic, and algebra. According to Mac Lane, Steenrod stated that the initial category paper "had more impact on him than any other research paper; other papers contributed results, while this paper changed his way of thinking."[83] As such, category theory is the Ur-example of the abstractionist epistemology that prevailed in mathematics during the postwar decades. When postwar mathematicians wrote that mathematics consisted of "abstraction upon abstraction," it is certain that they had category theory in mind, which took the concepts of abstract algebra and transformed them into the concrete subject matter of a new abstract theory.

In their 1945 article, Eilenberg and Mac Lane explained, "In a metamathematical sense our theory provides general concepts applicable to all branches of abstract mathematics, and so contributes to the current trend toward uniform treatment of different mathematical disciplines."[84] Even though the original impetus for the development of the category theory was a particular topological problem, Eilenberg and Mac Lane soon realized its potential and framed their work within a prevalent attitude of mathematical research at the time. Mathematicians adopted category theory as a conceptual language that brought together topology, algebra, logic, and geometry. It did function to some extent as a metamathematical theory, but it was not philosophical. The language of categories and functors, and the method of diagram chasing, pervaded much of postwar mathematics. In mathematics, at least, unification reoriented the discipline from the inside out.

The merits of unification through abstraction were not limited to the danger of increased specialization. Mathematicians praised the abstract theories they constructed for providing an economy of thought and a simplified view of past results. Perhaps more than anything else, over time abstraction came to denote a theory of knowledge, a belief that new mathematical knowledge was attained only via an ongoing and processual abstraction. As will become clear in later chapters, while most mathematicians did not contest the basic premise that abstraction is a route for mathematical creation, some did question how far the process should go and whether it should be countered by other modes of research.

There is nothing that makes an abstract unified theory inevitably "better" than a theory that is limited in scope. As some midcentury mathematicians noted, what is gained in breadth is lost in depth. By substituting three theorems with one, the intricate details unique to each of the original three are lost as the focus broadens to encompass only those characteristics which are common to all three situations. Yet the adherence to unification was so strong that all associated concomitants were taken as unquestionably good. For mathematicians, unification equaled progress. Unification was the aspiration, but its horizon was perpetually receding.

Rip Van Winkle

By the 1960s, the effort at unification caused a generational break in mathematics. At its core, the split was about the relative goals and merits of mathematicians' unification project. As Birkhoff noted in his 1942 review of Lefschetz's book, mathematicians understood unification and abstraction as a means to an end. Propelled by the example of modern algebra, mathematicians held that generalized theories offered an economy of thought and insights into problems that were otherwise too complex to solve. When Zariski praised Chevalley's abstractionist tendencies, he was quick to add, "In algebra, he is not a generalizer for generalization's sake, but rather he has a taste and preference for the difficult and deep problems of his fields."[85] Chevalley's generalist attitude served a purpose.

But by the 1960s, some members of the mathematical community began to feel that unification had become an end in itself. This was especially true of older mathematicians who, like Birkhoff, could in the 1940s still praise (albeit with reservations) mathematical abstraction. By the 1960s, these same mathematicians turned to criticizing. As mathematical theories grew increasingly abstract, determining the relative utility and inutility of a given mathematical theory became a personal value judgment. There was no agreed-upon metric by which to determine whether a new theory was "useful." The generation of mathematicians who had received their training during the interwar period adhered to the abstractionist spirit that dominated mathematics, but their own training was still rooted in a classical approach. This was not the case for many of the mathematicians who joined the field in the 1940s. This younger generation did not spend its days studying curve theory, working through one example after another as Steenrod had done. Rather, they were introduced to topology through Steenrod's and Eilenberg's work. Their path was thoroughly axiomatic.[86]

In 1974, seventy-eight-year-old Wilder commented on this transformation. The increasing power of modern mathematics, "as evidenced in its ability to solve problems hitherto inaccessible," was due in part to the forces of abstraction and generality. It was also due to the power of the younger mathematicians, "who enter a field at the high level of abstraction already attained when its capacity is nearing a maximum; to them, the high level of abstraction seems natural—it is where they *begin* their research (and, incidentally, where many of the oldest generation of mathematicians leave off, or turn to problems on levels at which they have been formerly successful)."[87] Wilder's last comment gestures toward the generational fissure that began to widen in the 1960s.[88]

The spirit of rewriting did not cease in the 1960s. Perhaps the most celebrated mathematician of the period, Alexander Grothendieck, was famous for creating abstract theories that led to the solutions of existing problems (often using category theory).[89] Grothendieck's metaphor of the "rising sea" symbolized this idea. Gothendieck held that there are two ways to crack a nut. The first is to hit it repeatedly from different angles until it opens. The second is to immerse it in water for an extended period of time until it opens to the touch. The same holds true for mathematical theorems, Grothendieck contended, and the latter method is the superior one. His aim was to construct an abstract theory ("the rising sea") so that the original problem would practically dissolve.[90] Grothendieck's theoretical work gained a large following, but some of the older generation were not so keen.

Take Marston Morse. In an article published in 1940, Morse celebrated the abstractionist tendencies of modern mathematics: "Popular opinion to the contrary, abstract studies are often simplest. The process of abstraction rules out the irrelevant and permits greater generality. Mathematicians abstract in order to unify, simplify, comprehend and extend."[91] Yet by 1970, he too had grown wary. In a letter to French mathematician Arnaud Denjoy, Morse conceded, "I agree with you that many of the young mathematicians are devising algebraic abstractions which are obvious when they are relevant, and in general take more space to explain abstractly than it takes to establish the desired theorems without their use."[92] The three decades that had passed had seen an unprecedented growth in the number of mathematics PhDs. The younger generation entered a field in which axiomatic theories and an emphasis on generalization were already taken as de facto mathematical values.[93]

When in 1933 Steenrod took Morse's class at Harvard, he complained that Morse was only interested in the combinatorial approach and had "no patience" for his geometric questions. A decade and a half later, it was Morse that felt out of touch with contemporary developments. When Morse first met mathematician Raoul Bott, who came to the IAS upon receiving his PhD in 1949, Morse was "in a sense a solitary figure, battling the *algebraic topology*, into which his beloved Analysis Situs had grown."[94] Bott remarked that "the development of the algebraic tools of topology, or the project of bringing order into the vast number of homology theories which had sprung up in the thirties—and which was eventually accomplished by the Eilenberg-Steenrod axioms—these had little interest for him."[95] The mathematicians who started their careers after World War II did not have such latitude. Studying topology meant mastering the homology theory constructed by Eilenberg and Steenrod. It was the language in which the conceptualization of the field was written.

The feeling that abstraction had gone too far was especially prevalent among older mathematicians who believed that the spirit of unification obscured any connection between the contemporary frame of a given theory and its particular origin. Older mathematicians held that the perpetual quest for increasing abstraction went so far as to completely obscure if not erase the *meaning* of the original theory. In 1964, Louis J. Mordell, a seventy-six-year-old British mathematician, published a scathing book review in the *Bulletin of the American Mathematical Society*. Serge Lang had authored *Diophantine Geometry* two years earlier. Mordell criticized Lang for a cumbersome presentation style, lack of clarity, flagrant disregard for detail, and confusing definitions. An interested reader, Mordell complained, would "require the patience of Job, the courage of Achilles, and the strength of Hercules to understand the proofs of some of the essential theorems."[96] Toward the end of his seven-page review, Mordell finally divulged why *Diophantine Geometry* had made him break out in critical hives. Mordell, who had founded the field decades earlier, was bewildered by what the field had become: "The reviewer is reminded," he exclaimed, "of Rip Van Winkle, who went to sleep for a hundred years and woke up to a state of affairs and a civilization (and perhaps language) completely different from that to which he had been accustomed. There were, however, some things still familiar to him—which is more than can be said by the reviewer about the presentation of the present treatment."[97] In Mordell's view it was not so much that the modernist approach erased the past, but that it had made it unrecognizable.

Upon reading the review, German mathematician Carl Ludwig Siegel, another early founder of the field who was then in his late sixties, immediately dispatched a letter to Mordell to sympathize. He wrote, "When I first saw the book, about a year ago, I was disgusted with the way in which my own contributions to the subject had been disfigured and made unintelligible. My feeling is very well expressed when you mention Rip van Winkle!"[98] Four years earlier, Siegel had been invited to spend the coming academic year as a visiting scholar at the IAS. Robert Oppenheimer, who at the time was the director of the IAS, sent Siegel an official invitation. Siegel immediately rejected it.[99] While in his letter to Oppenheimer he cited health concerns, he felt the need to write an additional, more personal, letter to his colleagues in the School of Mathematics (including Marston Morse, Oswald Veblen, Albert Einstein, and Kurt Gödel). "I was frightened," he wrote, "by the idea that my stay at the Institute for a longer period of time would not be welcomed by the younger generation of mathematicians whose scientific taste and interest seems to be very different from my own."[100] He then added, "It is quite possible that the *abstract dreams* of these people ultimately may lead again to fertile ground and to real progress. However, I feel unable to adapt myself to this modern spirit, and I prefer to continue my own work in the style of the 19th century."[101] Many of the mathematicians of the older generation, such as Siegel, idly observed the sort of abstract fashion that dominated the mathematical landscape around them, but they had no desire to participate.

To appreciate the remarkable transformation that mathematics underwent in the decade following World II, it is worth noting that in 1945, as the war was drawing to an end, the faculty of the School of Mathematics at the IAS had formally recommended an additional member to join their midst. Having considered several candidates, the faculty unanimously supported the very same Carl L. Siegel, who, they noted, "eclipsed" all other candidates. Siegel had earned his PhD in 1920 from Göttingen. A pacifist, he moved to the US and settled at the IAS in 1940. Now his colleagues wished he would stay. "There is no one in Siegel's generation and among the younger mathematicians," the faculty noted in their recommendation, "who is of comparable strength, and he is now at the height of his productivity."[102] Siegel was forty-eight years old at the time. Such a glowing assessment was confirmed by leading mathematicians from around the world. Godfrey H. Hardy of Cambridge University wrote, "I don't remember any time when anybody questioned that he was the equal of any mathematician of his generation," while Chevalley announced that Siegel's work thus far placed "him on the level with a Hilbert or a Poincaré."[103] Lefschetz wrote in that Siegel was "beyond any doubt the most gifted and original mathematician of his or lesser age anywhere," and Richard Courant, who knew Siegel from their days in Göttingen, similarly noted that Siegel was the strongest mathematician he had ever met.[104]

In describing his various contributions to mathematics, the official recommendation letter went into detail accounting for what made Siegel stand out among his contemporaries, and what characterized his unique research approach: "Siegel prefers to carve the stone of hard concrete problems rather than to knead the soft clay of general abstractions." It was not that he was unable to handle the "modern machinery of abstract concepts and methods," but when he did, he "use[d] them to solve some pre-existing concrete problem which had before been unapproachable."[105] By 1945, the mathematicians at the IAS recognized that Siegel was more "classically minded" than most of the mathematicians around him, but that did not stop them from recommending him to full professorship in the institute or from celebrating his work. It is not that the distinction between the concrete and the abstract did not already exist, but in 1945 Siegel could still find a home at the IAS. Fourteen years later, after he had returned to Göttingen to rebuild the mathematical community there, he no longer felt he could fit into the mathematical community around the institute.

To appreciate the feeling of estrangement felt by mathematicians of the prewar generation, it is worth returning to Rip Van Winkle, who upon meeting his own son, exclaimed: "I'm not myself—I'm somebody else—that's me yonder—no—that's somebody else, got into my shoes—I was myself last night, but I fell asleep on the mountain, and they've changed my gun, and everything's changed, and I'm changed, and I can't tell what's my name, or who I am!"[106] Rip van Winkle fell asleep for twenty years, which is approximately the time it took the mathematical landscape in the United States to transform.

Neither mathematical modernism nor unification nor abstraction had a clear end. The three were ongoing adjunct projects that entailed the writing and rewriting of mathematics as a body of knowledge. The ripple effects of this project were far-reaching. As will become clear in the following chapters, it redefined the relation of mathematics to the world, reconstructed mathematics as art, and even suggested a new historiography.

CHAPTER TWO

Applied Abstraction

Axiomatics and the Meaning of Mathematization

They [axioms] are no longer assertions about contents that have absolute certainty, whether it be conceived as purely intuitive or rational. They are devices which must be so broadly and inclusively conceived as to be open to every concrete application that one wishes to make of them in knowledge.
—Ernst Cassirer

In a 1946 memorandum, mathematician John L. Synge warned his colleagues, "There is a danger of misunderstanding as to the meaning of 'applied mathematics.'"[1] Synge had immigrated to the United States from Ireland just before the war. Now, at its end, he was charged by the mathematical community with chairing a committee to study the future of the field. The task of the committee was relatively simple: it had to determine whether a separate organization dedicated to applied mathematics should be founded within the already established American Mathematical Society (AMS). Yet before the committee's work could begin, Synge insisted that a better understanding of what applied mathematics *is* was in order. "I would value a statement as to what you understand by the term," he wrote to committee members.[2] Synge encouraged his fellow committee members, among whom were German émigré Richard Courant, polymath John von Neumann, and topologist turned statistician John Tukey, to forward his query to their colleagues at other institutions as well.

By November, Synge had received more than twenty-five letters from mathematicians working around the country. However, no consensus emerged.

In perhaps the most revealing response, John H. Curtiss, who had recently been appointed to head the newly established Division of Applied Mathematics at the National Bureau of Standards, wrote to Synge, "I have tried to do a little thinking lately on what applied mathematics really is. So far I have failed quite miserably to arrive at any conclusions."[3] Recognizing the ironic position he was in, Curtiss wryly commented, "It seems too bad to be founding an institution with such a name and not to be able to define what the name means."[4] Those who answered Synge's request could neither agree on what applied mathematics was nor on what its institutional relation to pure mathematics should be.[5] Why did mathematicians find it so hard to define what applied mathematics was?

Mathematicians' inability to define what applied mathematics was stemmed from a broad transformation in the meaning of *mathematization.* In the 1930s, mathematics began transforming from a tool associated most closely with measurement and calculation into an instrument for *theory construction.* Two developments precipitated this view: Hilbert's axiomatics and Einstein's theories of relativity. Coming close on one another's heels, together they persuaded mathematicians that a fundamental division existed between the analytic and phenomenological aspects of a theory. The difference between mathematical and physical concepts, mathematicians argued, was one of degree rather than kind. A point, a line, mass, energy, and time were all examples of abstract concepts defined logically, *not* inductively. In other words, mathematicians argued that while physical theories were rooted in observation and experimentation, their mathematical formulation must necessarily be independent of the lived world. Consequently, the way to ensure the veracity of a given theory was *not* to build it from the ground up by experimentally defining its basic concepts and relations and deriving and testing its consequences. Rather, since the analytic construction of a theory begins from above, only its theoretical deductions (and not its basic assumptions) could be subjected to external verification. During the war and the decades immediately following it, mathematicians insisted that the crucial role of the mathematician interested in worldly problems was, first and foremost, *formulation.* Defining the basic concepts of a theory and the relations between them was a mathematical act. Once this initial step was accomplished, the problem could be solved by any mathematically inclined individual, but it was not and should not be confused as the true task of the mathematician.

While this view of mathematization preceded World War II, the proliferation of new mathematical fields, from operations research to communication theory, computing, and, most crucially, game theory cemented this view of theoretical construction. From a historical vantage point all of these fields are distinct from one another and separate from mathematics, but in the aftermath of the war such distinctions could not be easily marked. They were mathematical theories, and as such they represented the unbounded potential of mathematical analysis, especially of the axiomatic method. More fundamentally, these sciences of war also extended the reach of mathematization as theory construction from its original basis in physics. As I show in chapter 3, nowhere was this transformation more evident than in the postwar social sciences. Just a couple of years before Synge's committee convened, von Neumann (himself a member of the committee) published the now famous *Theory of Games and Economic Behavior* in collaboration with Oskar Morgenstern. The book, which quickly gained the attention of mathematicians and economists, not only ushered game theory into the field of economics and the rest of the social sciences, but also (as will be made clear soon) redefined the nature of mathematical theory in explicating social phenomena. It was not measurement and quantification that characterized the mathematization after the war, but axiomatization.

Mathematization, thus, is a historical concept. As mathematics changed, so did what was entailed in applying it. The twentieth-century image of mathematics as both a body of knowledge and a discipline is so remarkably alien to the vision mathematicians held in the nineteenth century that it would be unrealistic to expect that what it meant to apply mathematics to a given phenomenon would remain the same.

Paul Erickson, Judy L. Klein, Lorraine Daston, Rebecca Lemov, Thomas Strum, and Michael Gordin have argued that game theory, operations research, and rational choice theory are indicative of what they term Cold War rationality. Cold War rationality, they write, was formal and algorithmic. It sought to break "complex tasks and episodes . . . into simple, sequential steps; the peculiarities of context, whether historical or cultural, gave way to across the board generalization; analysis took precedence over synthesis."[6] Erickson et al. argue that this form of rationality arose in reaction to the fear of atomic destruction, which compelled postwar intellectuals to separate rationality from reason and to formalize decision-making. Cold War rationality shares much with axiomatic reasoning. However, the two do not overlap. As the authors write, most of the components they attribute to Cold War rationality predate the Cold War, but "it was the Cold War that consolidated and glamorized them."[7] It seems unquestionable that the Cold War provided not only the institutional framework (the centers and the funding) but also the impetus for the development and adoption of game theory, operations theory, and allied fields. However, the history of axiomatic thought and the changing meaning of mathematization are necessary for understanding how this wholesale transformation in the meaning of rationality was even possible. The separation of formal analysis from observed phenomena, and the move from an inductive to a deductive conception of theory, was premised on the existing axiomatic conception of mathematics.

Erickson et al. write that, depending on the variant of Cold War rationality one seeks to understand, different "genealogies" are available: research into the logical foundation of mathematics, von Neumann and Morgenstern's *Theory of Games and Economic Behavior*, and the growth of applied mathematical fields in response to World War II and the Cold War.[8] What I wish to emphasize here is that all three "genealogies" they identify are axiomatic through and through. Research into the logical foundations gave rise to modern axiomatics á la Hilbert, and to the autonomous conception of analytic formalization. Game theory extended axiomatics beyond the confines of mathematics and physics and into the social sciences, demonstrating the benefits of analytic formalization. Finally, the growth of mathematical applications during the war and its aftermath was advanced by the work of many trained mathematicians who were steeped in axiomatic thinking. These mathematicians (Albert Tucker, Merrill Flood, Henry Wallman, Richard Bellman, John Curtiss, George Dantzig, and John Tukey, among others) did not shy away from the utility of mathematics, but they were trained to think about mathematics as a formal system and they brought this understanding with them to their work. In drawing attention to axiomatic thinking, my aim is twofold. First, it demonstrates that defense intellectuals had a reason to believe formal analysis was not only the appropriate response to the political moment but that it was justified on epistemological grounds. The intellectual justifications for abstract formalism were laid *before* World War II. Second, as I show in the following chapters, at least some postwar researchers held onto axiomatics because they believed it offered a powerful tool with which to uncover otherwise obscured truths. Some postwar researchers believed (together with mathematicians) that decontextual analysis was necessary not (only) because of the threat of nuclear war and the fallibility of human rationality, but because they held that certain truths can only be accessed by abstraction away from historical and social contexts.

Mathematics at War

In March 1942, the president of the American Mathematical Society, Marston Morse, and Marshall Stone, who the following year succeeded him as president, met in Washington, DC, with the chair of the National Defense Research Committee (NDRC), James Bryant Conant, the director of the Office of Scientific Research and Development (OSRD), Vannevar Bush, and the president of the National Academy of Sciences (NAS), Frank Jewett. Stone and Morse, representing the American mathematical community, had attempted for the previous two years to integrate mathematicians more fully into the country's scientific defense efforts, but felt that to date their endeavors had failed miserably. They requested the meeting with the three men in hopes that they could convince them to fully mobilize the mathematical talent of the United States in coordination with the Office of Scientific Research and Development. Stone and Morse considered the meeting to be a last-ditch effort. In a letter to Morse written seven months earlier, Stone fumed, "It is time for us to stop sitting quietly by while the physicists, chemists, and engineers monopolize the contributions to be made by the exact sciences under the OSRD. I am afraid that it may be necessary for us to exert real pressure to accomplish any change in the existing situation and to injure our modesty in the process. Nevertheless, it is my conviction that we should go ahead and do so without delay."[9] The meeting gave Stone and Morse an opportunity to present their case in person. In preparation, they drafted a memorandum entitled "Mathematics in War" to present to Bush, Conant, and Jewett, in which they explained their position in greater detail.[10]

Stone was not one to shy away from an argument. In December 1939, Stone had published an article in the *Nation* protesting the implementation of new tenure and employment guidelines instituted at Harvard, which placed a term limit of nine years for junior faculty. Announcing a "crisis at Harvard," Stone called out James Conant, the president of the university at the time, for not consulting the faculty when deciding on the new policy. He wrote, "The existence of such a crisis has therefore posed a fundamental constitutional question, one which may well prove, now or later, to be the most important point at issue: Can the complex modern university be governed both wisely and satisfactorily without effective, constitutional participation of its faculties in the decision of questions of general policy bearing directly on their several educational functions?"[11]

Stone's legalese originated in the nation's highest court. A year before the five men met in Washington, President Franklin D. Roosevelt had appointed Stone's father, Harlan F. Stone, as chief justice of the Supreme Court. While the younger Stone chose to pursue mathematics instead of following in his father's footsteps to become a lawyer, his father continued to send him his court opinions.

"With concern and regret," Stone and Morse's memorandum begins, "we express our opinion that the existing provisions for the employment of mathematicians in the professional service of their country are both inadequate and short-sighted as viewed against the realities of total war and its aftermath. We strongly urge the immediate alteration of prevailing arrangements so that the mathematical intelligence of the country can be brought fully to bear upon problems of war now confronting us."[12] In order to amend this state of affairs, the authors suggested that the OSRD should hire a designated mathematician who would serve as a liaison between the army and the mathematical community. This mathematician would be charged with analyzing defense research programs, determining the mathematical content and difficulty involved in any given problem, and assigning mathematicians to work on various projects accordingly. To fulfill this highly valuable job, the authors suggested that the desired individual must be well versed in modern pure mathematics and have high standing in the mathematical community and wide acquaintance with its members. To make sure their point came across clearly, the authors added, "It would be a serious mistake to demand the further and otherwise natural qualification of extensive experience in some fields of applied mathematics."[13] The authors insisted that only a pure mathematician could get the job done, both because there was no strong research tradition in applied mathematics from which to draw and because a broad view of the entire mathematical landscape was a necessary qualification.

The meeting lasted an hour. Morse and Stone did not achieve exactly what they wished, but their plea was acknowledged. Instead of a one-man liaison, it was agreed by all parties that an advisory committee of mathematicians would be appointed to mediate between mathematicians and the NDRC. The night after their meeting, Jewett penned a letter to Morse asking him to compile a list of possible nominees for the committee. Three days later, he followed his first letter with additional thoughts: "The more I think of it," Jewett wrote, "the more I feel that if the Mathematical Committee is to render a full measure of assistance, it should be not only eminent but should be composed of both fundamental and applied mathematicians."[14] In the three days that had passed, Jewett had carefully reviewed the memorandum Stone and Morse presented to him in the meeting. He later explained, "I had the impression that the authors were proposing a committee wholly of fundamental or 'pure' mathematicians. While I can hardly qualify as a mathematician in either category, I have seen enough of the problems both in the last and this war to feel that a committee which did not include some applied mathematicians would be less effective than one which did. As a matter of fact, I think it might lead rather directly to the organization of a group of applied mathematicians."[15] It took an additional month to form the committee, which, following Jewett's insistence, consisted of both pure and applied mathematicians.

Despite the best of intentions, the NAS-NRC (National Academy of Sciences and the National Research Council) committee, as it was known, had no significant impact on the war mobilization of American mathematicians. Nine months after the initial meeting, as part of the reorganization of the OSRD, Bush established the Applied Mathematics Panel as the central body in charge of directing mathematical work. The work of the NAS-NRC committee, in effect, became futile. As the panel's chair, Bush appointed Warren Weaver. An engineer by training, Weaver was well-known among professional mathematicians, but he represented exactly the type of mathematician Stone and Morse had advised against. Having insisted that it would be a grave mistake to let an applied mathematician run the mathematical community's war effort, Stone and Morse considered the decision to be nothing less than a slap in the face.[16]

In retrospect, it is not surprising that Stone and Morse's proposal failed. Their plan was antithetical to the collaborative ethos that characterized war-related research. Their scheme was based on the strength of the individual pure mathematician, not on the OSRD's ideal of interdisciplinary teams of scientists and engineers. It emphasized the theoretical aspect of the work, not the experimental and pragmatic parts.[17] Moreover, it insisted that the historical record made clear that a separation existed between the two. "A brief consideration of the relation of mathematics to the sciences," the memorandum began, was necessary in order to better understand the current situation.[18] The history of science, the authors explained, had proven time and again that mathematics had played a fundamental role in the discovery of natural laws. At times mathematics had proceeded in conjunction with physical investigation, but just as often—and this was key—mathematics developed without any direct relation to the physical reality proven most useful in physical theories. "It cannot be too strongly emphasized," they wrote, "that in this sense the scientific function of mathematics is a much broader one than that most often put to use by the physicist or the engineer—namely, the ancillary function of providing convenient solutions and means of computation for particular problems *already formulated and delimited by him in terms of those branches of mathematics with which he happened to be more or less familiar*."[19] Contrary to this popular (albeit mistaken) view, the role of mathematics, Stone and Morse insisted, was not limited to calculation. Much more fundamental was the role mathematics played in formulating problems in the first place, and it was in this task that mathematicians excelled.

A year earlier, in a long letter to mathematician Dunham Jackson, who had served with Stone on an AMS committee, Stone outlined his ideas in greater detail. His remarks highlight how irreconcilable his views were with those of the scientists in the OSRD. Stone conceived of the work of mathematicians as wholly and completely an individual affair: "A mathematician will undoubtedly work most efficiently under the conditions to which he has already adapted himself."[20] Thus, he explained, research mathematicians should remain in their academic position and simply swap their own research work for their assigned problems. For Stone, the mathematical content of the problem existed (or at least potentially could exist) completely separate from its physical origin. Considering that some of the problems that might arise during the war would be theoretically advanced, Stone suggested that the best way to ensure a quick solution would be to place the problem before mathematicians as a challenge. He wrote, "It should be possible to divest such problems, once recognized, of their military character and to set them as public 'prize' problems."[21] Stone firmly believed that familiarity with the context from which a theoretical problem emerged was not necessary for solving it.

Stone's views may have sounded outlandish to the physicists in the OSRD, but they reflected mathematicians' ideas about mathematics' role in physical theories. Stone partially acknowledged that fact when he noted in his letter to Jackson that if they were to succeed, mathematicians would have to engage in a bit of propaganda. The greatest limitation facing mathematicians, according to Stone, was "the general American attitude of antagonism to theory in general and to mathematical refinements in particular and the abysmal ignorance of the majority of intelligent Americans concerning the uses of mathematics."[22] To understand how Stone and Morse arrived at this view of the relation of mathematics to physics, it is necessary to go back in time to the generation of American mathematicians that had come before them. This conception of mathematics and physics seemed like common sense to American mathematicians in the two decades before the war. The catalyst was, as with many other things, the discovery of Einstein's general relativity.

Veblen and Birkhoff

Both Stone and Morse earned their PhDs under the supervision of George Birkhoff at Harvard University. Birkhoff was one of the first American-trained mathematicians to receive international recognition during the first decades of the twentieth century. Morse later recalled that, unlike some of his American counterparts, Birkhoff "thought of his contemporaries in Europe, especially in Germany, as colleagues rather than teachers."[23] Birkhoff settled at Harvard early in his career and remained loyal to the university until his death in 1946. He helped transform the Department of Mathematics from a service department to a well-recognized research center, and many of his students had successful careers in mathematical research. According to Morse, Birkhoff adhered to a New England ideal of self-sufficiency, and carried an air of detachment that was often mistaken for disinterestedness.

A student of E. H. Moore in Chicago, Birkhoff was a contemporary of Oswald Veblen, who for his dissertation constructed a new axiomatic system for Euclidean geometry, which required only two rather than three undefined elements.[24] His work was part of American mathematicians' general interest in studying how different axiomatic systems could be applied to the same underlying phenomena, which turned axiomatic analysis into a subject in its own right.[25] For them, abstract mathematics was axiomatic analysis. In 1905, at the invitation of Woodrow Wilson, Veblen took a position at Princeton University, where he influenced a generation of young mathematicians.[26] Princeton's centrality to both the national and international mathematical community in the 1930s was due in no small part to his work.

Veblen was especially influential in popularizing Hilbert's axiomatic approach among young mathematicians. In 1910, together with his student John Wesley Young, Veblen published the first of a two-volume treatise entitled *Projective Geometry*. The book, which was intended to be a textbook for graduate students, was written explicitly in Hilbert's style, and as such served as a clear and definitive introduction not only to projective geometry but also to the modern axiomatic method. Unlike Hilbert, Veblen and Young begin their book with a step-by-step introduction to axiomatic thinking. The choice of postulates, they explain, is "in no way prescribed a priori, but, on the contrary, is very arbitrary."[27] They add that "the words point and line are to be regarded as mere symbols devoid of all content except as implied in the assumption concerning them."[28] Once they introduce the reader to the modern axiomatic method, the authors construct an axiomatic system for projective geometry.

Birkhoff and Veblen kept in touch throughout their careers. They were united in their deep devotion to the advancement of mathematics in the United States.[29] Each published textbooks, trained students, edited prestigious mathematical journals, and advocated for mathematics on the national stage. They did not always agree on how best to accomplish their goal, as, for example, in the 1930s, when Birkhoff believed that finding jobs for American-trained mathematicians was their first priority and did not support Veblen's admirable efforts on behalf of European mathematicians. However, their dedication to the advancement of mathematical research cannot be overemphasized. Abraham Taub later remarked that, together with Gilbert Bliss in Chicago, Veblen and Birkhoff "determined everything about American mathematics" by deciding who got the best jobs and who received fellowships from the National Research Council.[30]

In 1925, when Stone was twenty-three years old and completing his dissertation, Birkhoff advised Veblen as to how to recruit Stone to Princeton. Praising Stone for his ability and determination, Birkhoff reported that on a recent visit to Washington he had met with Stone's father, who a week earlier had been nominated by President Calvin Coolidge to the Supreme Court. The elder Stone confirmed his son's desire to come to Princeton. However, Columbia had already offered Stone a position with twelve hours of teaching. Birkhoff added, "His father and I were agreed that it would be undesirable for Stone to teach more than 9 hours."[31] Therefore, if Veblen wished to make an attractive offer to Stone, Birkhoff advised, he should reduce his teaching load: "You ask about his scientific standing in comparison with other men of his generation. It is a little too early to be sure, for Stone is so exceedingly young. I should expect him to go a long way."[32] Seven years later, Stone returned to Harvard to become Birkhoff's colleague. By 1937, he had advanced to a full professorship. His relationship with Birkhoff remained close throughout.

Both Birkhoff and Veblen began their careers as pure mathematicians,

Birkhoff specializing in analysis and Veblen in geometry. However, during the 1920s they both found themselves increasingly interested in physics. Einstein's discovery of general relativity made a strong impression on each of them, and both began publishing on the topic. In addition to igniting their interest in contemporary physics and altering their respective research agendas, relativity also made each man reexamine and reevaluate the relation of mathematics to reality. Was mathematics a tool for discovery? Was it a heuristic? Was it simply a descriptive language? Or was it a deductive tool that enabled scientists to more efficiently predict experimental results?[33]

Veblen's and Birkhoff's training in the abstract point of view that characterized American mathematics at the beginning of the century deeply informed their study of relativity and convinced them that a clear separation existed between the analytic and the phenomenological parts of a theory. Veblen neatly captured this view when he noted that "the abstract mathematical theory has an independent, if lonely, existence of its own."[34] It is a belief in this fundamental partition that helps explain Stone and Morse's memorandum.

Birkhoff first published on general relativity in 1922, when he reviewed several recently published books and articles written on the theory of relativity. He followed this a year later with his own book, titled *Relativity and Modern Physics*, which served as an introduction to relativity geared toward undergraduate students.[35] That same year he also devoted his six public Lowell lectures to the theory of relativity. The lectures, which were later published in a book, *The Origin, Nature, and Influence of Relativity*, most clearly outline Birkhoff's ideas about the relation between mathematics and reality. He explained, "There is . . . a sense in which relativity can be made to throw further light upon the mathematical-logical factor in mind and nature, namely, as a typical instance of the abstraction."[36] For Birkhoff, the greatest lesson furnished by relativity was the increased scrutiny it demanded of the fundamental terms physicists (and, more generally, scientists) employed in their theories.

Einstein's theory of general relativity challenged the idea that time and space can be analyzed separately, arguing instead for a single space-time continuum. Following relativity theory, gravitational force, which determines the attraction between bodies, was nothing more than a curvature in space-time (represented as a four-dimensional manifold). Thus, the theory unseated the centuries-old conception of absolute time and absolute space and called into question any attempt to construct a physical theory based on naive correspondence between observation and reality. Birkhoff believed that now more than ever scientists must become more vigilant and explicit about the basic theoretical concepts they employed in their work. "Because the familiar classical abstraction of space and time [had] been held sacrosanct and qualitatively different from all other abstractions," Birkhoff explained, their redefinition could shed light on the nature of scientific abstraction more broadly.[37] The crisis of referentiality brought on by Einstein's relativity inspired Birkhoff and other mathematicians to reexamine the way in which mathematics applies to the world.

As Jimena Canales demonstrates, not everyone in the philosophical and scientific community was as quick to accept the implication of Einstein's theory. For Henri Bergson in particular, Einstein's claim that "the time of the philosophers does not exist" supplied inexhaustible fuel for his critique of relativity theory.[38] However, if philosophers were least amenable to the radical reconception of time implied by Einsteinian relativity, then Birkhoff and Veblen were the perfect audience. The impulse to question and overhaul science's most basic assumptions must have felt completely natural to these two mathematicians who were well versed in the foundational concerns that had lately captured the attention of the mathematical community. It is no great surprise, then, that both Birkhoff and Veblen were open to accepting Einstein's abstract conception of time, for only a few years earlier they had concluded, following Hilbert, that even a point and a line, concepts much simpler than time, could not be adequately defined by induction.

As far as both Birkhoff and Veblen were concerned, the theory of relativity did not imply that a more thorough investigation of either the psychological or philosophical aspects of time was in order. Rather, it persuaded them that the axiomatic understanding of time should be extended to all physical concepts. In a 1922 address entitled "Geometry and Physics," Veblen made the case: "The old way of accounting for this difference was to say that the electron is a substantial object whereas the point is only an abstraction. This way of dismissing the question will not satisfy us to-day, for we believe that the electron and the point are both abstractions. Moreover, the difference which we are seeking to explain is one of degree rather than of kind."[39] Veblen brought with him to his study of relativity the same rationale that had dominated his thinking since his student days. In the same way that concepts such as points and lines make sense only within a formal axiomatic system, so, too, did physical concepts such as time, energy, and mass.

While both Veblen and Birkhoff believed that all abstract concepts originate in experience, they insisted that their definition ultimately depended only on logic and that their mathematical formulation was *essentially* independent. Birkhoff, who favored a more philosophical approach and was influenced by Alfred North Whitehead, formulated his own theory of abstraction.[40] In his Lowell lectures, Birkhoff outlined his abstractionist theory using time as an example. In order to arrive at the abstract concept of an "instant of time," Birkhoff explained, one begins by observing that events occur in chronological order. Noting that event A happens before B, and similarly that B happens before C, the researcher may begin, Birkhoff continued, by postulating that this implies that A happens before C (i.e., they obey a given order). In the next stage of systematic elaboration, the events are numbered, and arithmetic is used to analyze them. It is only in the third and last symbolic stage that the idea of an "instant of time" is achieved through the abstraction of the time continuum. This symbolic stage can then be rigorously established by approaching "instant" as an undefined term, and "before" as an undefined relation.[41] "The final abstraction seems to stand out and exist at the mathematical level, apart from any concrete embodiment of it, in a very remarkable way."[42] Such claims about the independent nature of mathematics were different from the ones made regarding the autonomy of pure mathematics. It was one thing to claim that pure mathematical theories were independent of physical theories; what both Veblen and Birkhoff claimed was that even the mathematical theories of *physical phenomena* had a separate and autonomous existence.

In separating mathematical theory from phenomena, its actualization became subject to interpretation. In its austere mathematical formulation, the terms of the theory held no meaning. Regardless of the intuitive meaning inherent in their names, they were nothing but empty signifiers. Only once scientists attached a direct physical interpretation to the fundamental terms did the mathematical theory become a physical theory. Following Birkhoff's example of time, the concept of an "instant of time" had a physical meaning *not* because Birkhoff began his analysis by observing events in the world, but because after it was defined analytically it was reinterpreted as describing a physical phenomenon. As Veblen explained, "The places at which this life-blood of human meaning flows into the mathematical theory should, it seems to me, be the undefined terms."[43] However, once the relation between phenomena and mathematics was inverted—once it was no longer assumed that a physical theory arose directly out of reality—the door opened up to multiple interpretations. The same mathematical theory, if general enough, could account for more than one phenomenological context. All that was necessary was to provide a different interpretation to the undefined terms such that their new realization would still accord with a given natural phenomenon. For Veblen, this ontological multiplicity effectively distinguished the abstractions of geometry from those of physics (as he said, the difference was one of degree, not kind). Analytically the terms were defined implicitly by the relations set in the axioms. Since the axioms of geometry were less restrictive than those of a given physical theory, the undefined term (e.g., "point") could be given numerous interpretations, while a formalization of a physical theory with an undefined term being the "electron" was open for a more limited number of interpretations.

Moreover, declaring a clear separation between analytic formulation and physical reality implied that the verification was a post-theoretical endeavor. The way to judge whether a theory was true or not was not to test its axioms, which by definition were arbitrary and did not claim to capture unquestionable truths, but to check whether its deductions could be experimentally verified. The axioms, regardless of a theory's success, were provisional. In several lectures given during the 1920s and 1930s, Veblen returned to this point again and again. In 1923, Veblen railed against the teaching of geometry as a system of a priori truth: "A science resting on such supernatural basis was fittingly taught by the method of dogmatic indoctrination."[44] He did not object to the teaching of Euclidean geometry as a whole but wanted to ensure that its validity and significance were not taken as unquestioned truths. Veblen's ideas were not limited to Euclidean geometry or mathematics but extended directly into physics. "We have seen that an examination of the physical basis for our assumptions induces anything but a dogmatic frame of mind," Veblen insisted.[45] Thus, axiomatics expanded the role of mathematics. Besides providing scientists with a convenient language with which to numerically analyze phenomena, mathematics played an even more important role in safeguarding scientists against unconscious indoctrination.

Birkhoff expressed similar ideas when he appealed to abstraction as a way of striking a middle ground between radical empiricism and fully committed idealism. On the one hand, he disagreed with the operational view of his Harvard colleague Percy Bridgman, who insisted that only those concepts that could be described through physical actions had a place in theory. For mathematicians trained in the "abstract point of view," Birkhoff explained, attributing a reality to abstract concepts was absolutely natural. On the other hand, he did not believe, as some idealists might have, that there existed one abstraction that would eventually prove to be final and absolute. "Any abstraction serves only limited specific ends," he wrote.[46] We must understand, Birkhoff proclaimed, "that our only approach to a better understanding of the world is by means of a widening succession of abstract ideas, each explaining imperfectly some aspect of the stupendous whole."[47] Abstractions were real, but in a limited pragmatic sense.

By the late 1930s, developments in physics had convinced Birkhoff that the role mathematicians played in the advancement of science was even more crucial than they had realized. The dominance of theoretical physics made it apparent that mathematicians trained and well versed in abstractions had much to contribute. The situation in physics in the past few decades, Birkhoff explained in 1938, had "led naturally enough to the feeling that pure mathematics almost suffices without much recourse to the results obtained in the physical laboratory."[48] The separation between theory—understood as mathematics—and experimentation had become a truism for American mathematicians.

Developments in physics served as the impetus for this reexamination of the role of mathematics. However, mathematicians' conclusions were in no way limited to physics. *Axiomatics and relativity together redefined mathematization.* They provided a framework with which to understand the role of mathematicians in formulating theoretical knowledge broadly; not limited to measurement and calculation, mathematicians were also necessary for formulation. This was especially clear in Birkhoff's writings, as he extended his analysis to other sciences from psychology and biology to the social sciences. His belief in the potential of systematic mathematization cannot be overstated. In 1933, Birkhoff turned his critical attention to aesthetics. In a book-length monograph, Birkhoff constructed a mathematical theory of aesthetics that, he contended, could be applied to anything from the shape of the most basic polygons to intricate geometrical patterns, music, and poetry.[49] A few years later, satisfied with the success of this endeavor, he resolved to devise a mathematical theory of ethics.[50] "The world is of such highly technical, lawful structure," Birkhoff concluded in a public talk in 1937, "that we cannot understand it otherwise save by means of profound abstract thought, failing which the individual and even society itself can scarcely survive."[51]

Birkhoff, Veblen, and other mathematicians were, of course, not alone in reassessing the relation between mathematics and physical theories in the aftermath of Einstein's general relativity. According to Leo Corry, over the course of his career Einstein changed his opinion regarding the role of mathematics in physics. While in his early years Einstein considered mathematics to be merely a tool for physical intuition, later in his career he entrusted to it a more pivotal role as a source of scientific creativity.[52] In a 1933 lecture entitled "The Methods of Theoretical Physics," Einstein remarked, "I am convinced that we can discover by means of pure mathematical constructions the concepts and the laws connecting them with each other, which furnish the key to understanding natural phenomena."[53] Furthermore, as Gerald Holton has shown, Einstein believed that neither the fundamental concepts of physics (e.g., time and space) nor the axioms connecting them could be determined purely through induction. "Failure to understand this fact," Einstein wrote, "constituted the basic philosophical error of so many investigators of the nineteenth century."[54] As such, Einstein distanced himself from Machian philosophy and afforded a much more expansive role for mathematical investigation in the discovery of physical laws. In his autobiographical essay, Einstein, like many of the mathematicians of his generation and his colleagues at Princeton, pointed to non-Euclidean geometry as an important precedent for his thinking.

While Einstein and several other physicists began to question the conventional model of theory construction, it was the logical empiricists that turned their full attention to the "philosophical error" Einstein identified. As Alan Richardson notes, relativity "was a watershed for the new logical empiricist philosophy of science."[55] Whereas their goal was to assert the primacy of observation in physical theories, they now had to account for the noninductive nature of physical concepts and the increased role of mathematics in theory construction. A clear example is Rudolf Carnap's "Foundations of Logic and Mathematics," which appeared in 1939 as part of the International Encyclopedia of Unified Science. While the title might suggest that Carnap's topic was an investigation of the various philosophical schools that emerged in response to the foundational crisis in mathematics, this was not the case.[56] "It is one of the chief tasks of this essay," Carnap explained, "to make clear the role of logic and mathematics as applied in empirical science."[57] It was an analysis of the relation between theory construction and experimentation that Carnap was after, not the philosophy of mathematics. Carnap went to great lengths to explain the distinctions between pragmatic, semantic, and syntactic functions of language. However, the thrust of his essay was directed toward explaining the way in which a purely mathematical theory could be reinterpreted as physical.

Here, Carnap offered a solution to the new model of theory construction suggested by relativity. A mathematical theory is turned into physical calculi, Carnap explained, by providing semantical interpretations to its sign. Whereas the positivistic ideal required that the most elementary terms of the calculi would be interpreted first and that the more abstract terms would follow from the simpler ones, Carnap acknowledged that such a procedure did not accurately describe the work of theoretical physicists. The "direction in which physicists have been striving with remarkable success, especially in the past few decades," he wrote, has been the opposite (see Fig. 2.1).[58] Interpretation was provided first *not* to the least abstract but to the most abstract terms: "This method begins at the top of the system, so to speak, and then goes down to lower and lower levels. It consists in taking a few abstract terms as primitive signs and a few fundamental laws of great generality as axioms. Then further terms, less abstract, and finally elementary ones, are to be introduced by definitions."[59] Carnap went into great detail explaining how meaning was invested in the system from the top down, but the key was that his procedure offered a solution that maintained the primacy of observation while simultaneously accounting for the noninductive nature of physics' primitive concepts.

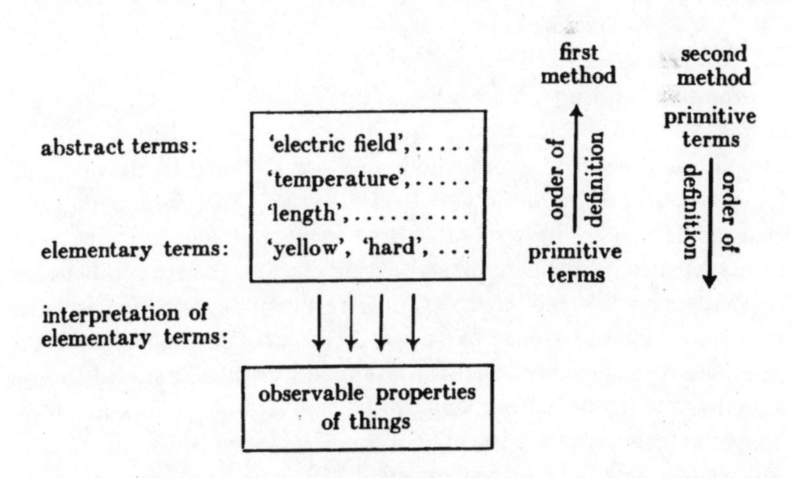

FIGURE 2.1 A diagram from Carnap's book illustrating the change from first defining elementary terms to first defining the abstract terms of a theory.

Source: Rudolf Carnap, *Foundations of Logic and Mathematics*

Instead of arising out of observations, mathematics now entered physics from the top.[60] Meaning was no longer predicated solely on the association of basic terms, such as points and lines, with sense experience. Instead, abstract mathematical concepts which were developed autonomously could be accorded with physical interpretations as long as the overall theory could be subject to experimental verification. Carnap's views were similar to those of Birkhoff and Veblen and arose from the same concerns, but logical positivists approached the problem from a different perspective than mathematicians. For Veblen, at stake was not philosophy, but practice. Claiming that an electron is an abstraction in the same way that a point is was something that mathematicians had emphasized in their training. Veblen popularized axiomatics in several of his publications and textbooks and insisted that this was how students should approach mathematics. Moreover, like Hilbert, he called upon his fellow mathematicians to turn their attention to physics and try to axiomatize mechanics and kinematics.[61]

Thus, while the writings of philosophers like Carnap have received much more scholarly attention than those of mathematicians like Veblen and Birkhoff, the latter were responsible for training the generation of mathematicians who found themselves involved in war-related research during World War II. These young mathematicians transformed a philosophical idea about the relation between theory and observation into new research projects.

Axiomatic Thinking

Despite their genuine efforts, Stone and Morse failed to convince the scientists at the helm of the OSRD of the crucial role pure mathematicians could play in the war. After Bush appointed Weaver to the head of the Applied Mathematics Panel during the war, the tension between the mathematicians and the OSRD only heightened.[62] Stone in particular took it upon himself to continue to fight on behalf of pure mathematicians. In a long correspondence that was eventually circulated among several scientists and mathematicians, Stone and Weaver argued back and forth about the "proper" utilization of mathematical talent in the United States. Much was at stake in their disagreement: who could speak for and represent the mathematical community and who was most qualified to develop mathematical applications for the war. With his engineering background,

Weaver did not adhere to the same theoretical conception of mathematics that Stone and Morse did.[63] Applied mathematicians, he insisted, were distinct from pure mathematicians in their training and scientific approach. What the war required, Weaver insisted in his letters to Stone, was down-to-earth practical mathematicians who were unselfish and capable of collaborating with others. These conditions, he explained, "exclude[d] a good many mathematicians,—the dreamy moon-children, the prima donnas, the a-social genius."[64]

Needless to say, Stone disagreed. Mathematicians were not technicians, he retorted. "When you know exactly what your problems are and have the appropriately trained technicians, the obvious move is to put technicians on the problem," he wrote.[65] However, Stone insisted, most defense-related problems were not clearly formulated. "It is in this situation," he said, "that quality of mind rather than specific training is the key to progress."[66] Stone, like his adviser, conceived of mathematics as a mode of thinking, a cognitive activity whose strength arose not from its particularity but its generality. He also held that a mathematical perspective could safeguard against dogmatism. "There is a good reason to believe," he wrote, "that reliance primarily on purely technical background may be dangerous even in problems which appear to be perfectly definite and straightforward. There is always advantage in getting a fresh point of view at least."[67]

Stone's and Weaver's positions represented two competing views of the proper development of applied mathematics after the war: one stressed the theoretical aspects of applied mathematics and the other focused on its pragmatic engineering utility. Considering the many mathematical subdisciplines that proliferated after the war, Weaver's view seems to have triumphed. However, the theoretical conception of the field did not diminish. Once the war was over, applied mathematics was no longer in the hands of scientists at the OSRD, but of mathematicians themselves, many of whom were first trained in pure mathematics. These mathematicians were able to find a compromise between Stone's and Weaver's positions. They adhered to the interdisciplinary and collaborative ethos of the postwar sciences, while advancing the idea that it was *mathematical thinking* rather than any particular mathematical theory that accounted for the utility of mathematics. For these mathematicians, axiomatic thinking and the autonomy of analytical formalization above all accounted for the promise of mathematics. A decisive influence in this regard was von Neumann's axiomatic development of game theory.

Much has been written about the development of game theory and its incredible effect on the social sciences in the postwar decades.[68] First published in 1946, *Theory of Games and Economic Behavior* was the work of polymath John von Neumann and economist Oskar Morgenstern. The book suggests that a multitude of economic scenarios could be described formally as games in which individuals seek to maximize their respective utility. While it began as an economic theory, by the 1950s and 1960s, game theorists sought to influence nuclear strategy, arms control, social psychology, political science, and evolutionary biology.[69] RAND scientists, who adopted and developed game theory early on, played a crucial role in disseminating it beyond mathematics and economics. In the process, they redefined nothing less than Cold War rationality.[70]

Whereas histories of game theory have primarily focused on the theory's impact on economic and national policy, less attention has been paid to the model of *mathematization* that game theory popularized. In her comprehensive study of rational choice theory, S. M. Amadae notes that game theory was the "catalyst for the novel axiomatic approach to human rationality" that prevailed at midcentury.[71] However, what were the underlying assumptions entailed in adopting an axiomatic approach? And what image of mathematical theory did the book advance? The axiomatic vision von Neumann and Morgenstern articulated in their book was demonstrably not Euclidean. Rather, the book tested Hilbert's vision, whereby mathematics was used not as a descriptive tool but an analytic one. Moreover, like Veblen and Birkhoff, von Neumann and Morgenstern insisted that mathematical theory existed in complete separation from physical phenomena. Game theory's immense success helped propagate this vision of mathematization among a broader audience that did not necessarily come from mathematics. Thus, beyond impacting military planning, political theory, and numerous disciplines, axiomatic game theory also helped redefine the role mathematics could play in explicating real-world phenomena.

Von Neumann and Morgenstern recognized that their conception of mathematization was novel and thus began the book with an extended meditation on "the role which mathematics may take" in the development of economic theory. They explained that, more than simply appropriating a new mathematical tool, such as functional analysis (although this too was the case), what they advocated was a radical reorientation of *how* mathematics impacts economic theory. The role of mathematics, they explained, was not to measure and compute, but to help formulate theories from the ground up by clearly defining the basic concepts of those theories and determining the underlying relations that obtain between them. In what would become a rallying cry for applied mathematicians in the postwar decades, the authors insisted that "there [was] no point in using exact methods where there [was] no clarity in the concepts and issues to which they [were] applied."[72] The authors explained that the first step in the successful use of mathematics was the formulation of the underlying problem in a way that would permit mathematical analysis.

In 1933, Von Neumann immigrated to the United States and joined the recently established IAS, where two of his colleagues were Veblen and Morse. At the time he was already celebrated by both mathematicians and physicists for his numerous contributions to both fields. During World War II, von Neumann was enlisted in the war effort, and it was while working on the Manhattan Project that he began collaborating with Morgenstern, a fellow émigré to Princeton. Von Neumann, who had spent time with Hilbert in Göttingen, turned his attention to axiomatics early on in his career. He first axiomatized set theory, and later did the same for quantum mechanics.[73] It is unsurprising that he brought the same arsenal to bear on economic theory.

Importantly, von Neumann broke with Hilbert. For Hilbert, axiomatization was done once a given science achieved a certain level of maturity. The goal of axiomatics was to clarify the fundamental assumptions and logical structure of an already existing theory. With game theory, von Neumann inverted the order: he used axiomatics to build a new theory from the ground up. In this regard, his approach hewed more closely to that of high modernist pure mathematicians who adopted axiomatics as a research tool rather than a philosophical one.

The authors' attention to the process of mathematization was not limited to the first chapter of the book. Besides as an introduction to game theory, the book can also be read as a careful guide to axiomatic reasoning. The book offered concrete step-by-step demonstrations of how this process proceeded in practice. The authors explained that the "mathematically rigorous and conceptually general" theory proceeded from a first heuristic step that included the "transition from unmathematical plausibility considerations to the formal procedure of mathematics."[74] They then, like Birkhoff did with an "instant of time," modeled that process in their presentation. They first outlined the overall basic economic situations motivating their theory, then showed how it could be translated into mathematical notions, and, finally, constructed a formal axiomatic system.

While the authors went to great lengths to explain how intuitive ideas were transformed into formal mathematics, they insisted that any formal mathematical presentation exists *independently* of intuition. As they phrased it immediately following an outline of the axiomatic presentation of a game, "This definition should be viewed primarily in the spirit of modern axiomatic method. We have avoided giving names to the mathematical concepts introduced in (10:A:a)–(10:A:h) above, in order to establish *no correlation* with any meaning which the verbal association of names may suggest." They added, "In this absolute 'purity' these concepts can then be the objects of an exact mathematical investigation."[75] This procedure, they explained, was the only way to ensure that basic concepts were clearly defined. Only once this fundamental break was acknowledged and understood could readers remind themselves of the "detailed empirical discussion" in the preceding sections and replace some of the mathematical symbolism with descriptive language.

This might sound at first like nothing more than language games, a simple act of translation whereby words are first replaced by mathematical symbols only to be reinterpreted later as the exact same words. Such an interpretation would miss the point. The exact mathematical formulation is essentially independent. Once constructed with the utmost rigor, the formal system sheds any bond to the empirical phenomena from which it arises. It is not, and should not be, thought of as a translation. Rather, it is an ad hoc construction. This is why game theory was not a theory that arose from specific contexts, but instead was later applied to such contexts. This notion of mathematization provided scientists with new analytic freedom that was especially potent in the social sciences.

To define the concept of utility, von Neumann and Morgenstern did not need to point to a particular behavior in the world, but merely show that the abstractly constructed concept was useful in analyzing specific situations. In the same way that a "point" had many interpretations, so did "utility." Indeed, in the same way that mathematicians defined a point *implicitly* as that concept that obeyed a certain set of relations, so did von Neumann and Morgenstern define utility. They wrote, "We have practically defined numerical utility as being *that thing* for which the calculus of mathematical expectation is legitimate."[76] In its original formulation, game theory (like geometry) existed simultaneously as both a self-contained mathematical system and a theory of economics.

Game theory was not the only new research area that challenged the classical conception of applied mathematics. Operations research, information theory, and computing similarly extended the bounds of mathematical theory beyond the classical areas of mechanics and mathematical physics. These new fields were not joined by a subject matter or by a common set of tools, but by a vague sense of shared approach. Together they redefined what mathematization meant. When in 1948 Claude Shannon published "A Mathematical Theory of Communication," he explained that he did not aim to present the mathematical formulation of the problem in its most general form or according to the highest "rigor of pure mathematics." However, he added that "a preliminary study . . . indicated that the theory can be formulated in a completely axiomatic and rigorous manner."[77] Shannon's remarks made clear that even mathematical theories that were not given axiomatic treatment were nonetheless influenced by it.[78]

The widespread confusion regarding applied mathematics in the aftermath of the war can mostly be blamed on the murkiness of the term itself, as mathematicians were not clear at first what distinguished mathematization from applied mathematics. By the late 1960s, it would be common to distinguish between applied mathematics and mathematical applications. However, throughout the 1950s such distinctions grew hazier as game theory, operations research, computing, and information theory began to be recognized as independent fields of research. For a while, axiomatization was the glue that cemented it all together.

Formulation

In 1956, ten years after Synge's committee struggled to define applied mathematics, the mathematical community again bent its efforts to this problem, this time on a national level. A report presented to the Division of Mathematics at the National Research Council began by posing the question, "What is applied mathematics?" It then asserts that no answer exists: "The Survey is emphatically not concerned" with a segment of mathematics or any grouping of subfields that could be singled out as *the* subject matter of applied mathematics. It aims instead at an activity that "consist[s] in the creation, the adaptation and the communication of mathematics inspired by and knowingly related to the efforts of advancing our *rational* understanding of some aspect of the world around us."[79] This broad-brush definition of applied mathematics as a sort of mathematics with an (albeit vague) purpose was common at the time. It enabled mathematicians to unify a collection of fields that were otherwise haphazard.

As elaborated upon in chapter 5, the purpose of the survey was to study what could be done to improve the state of applied mathematical research in the United States, and it gave rise to a temporary schism between pure and applied mathematicians. Given this political context, it is remarkable that in accounting for the current state of applied mathematics, the report nonetheless emphasized modern axiomatics: "Instead of external *a priori* truths," the report asserted, "the best that can be expected of them [axiomatic systems] is consistence—and the *a posteriori* satisfaction if they turn out to do their job well."[80] The implications of this transformation, the report announced, had been far-reaching. No field of pure mathematics was left unchanged, and applied mathematics would follow suit. "The release of mathematical creativity, made possible by the axiomatic approach, has given to applied mathematics a comparable opportunity of playing a more creative role."[81] Modern mathematics provided the applied mathematician with new theories and tools at his disposal, but more importantly, it "broadened his proper area of activity."[82]

When Veblen and Birkhoff celebrated the rise of modern axiomatics as a tool for the elucidation of physical theories, all the sciences of war were yet to be invented. Three decades later, after a host of new fields had proliferated, applied mathematicians still adhered to the same rationale. If anything, these new fields would ultimately vindicate the axiomatic method. The entire report, which covered every possible field of applied mathematics from the theory of automata to control theory and hydrodynamics, was permeated by the language of axiomatics. The axiomatic description is not limited to the natural sciences; it extended to engineering and the social sciences as well. By the mid-1950s, modeling had become the dominant feature of mathematical analysis across a variety of fields, and the report presented modeling as merely a natural extension of the axiomatic conception of physical theories. The mode of reasoning and the role played by mathematics, for these authors, remained remarkably similar.

Describing the use of mathematics in engineering systems such as communications engineering, the report offered the following florid exultation: "*The applied mathematician is expected to return from his sally into what, to the engineer, will always be the jungle of mathematics with a solid mathematical model in the bag.* A sense of engineering utility and feasibility provides the aim, and the practically infinite foresight of mathematical argument projects the bullet; the animal itself is a creature of the imagination with which the axiomatic approach has so plentifully stocked the woods of mathematics."[83] It is fair to assume that in its extended analogy, the report's meaning here was lost. Most simply, the passage asserts that the axioms constituting the foundation of mathematical models used in engineering systems, as in physical theories, laid no claim to capturing unquestioned truths. They were "creatures of the imagination," even though they were motivated simply by pragmatic utility.

Communication engineering, the mathematical theory of automata, game theory, operations research, and numerical analysis were presented as fulfilling the same ideal of axiomatic analysis. The report acknowledged, for example, that providing a precise definition of operations research might be impossible, but concluded that there, "as in all the previous contexts, the role of mathematics [was] primarily that of establishing a postulational model for the quantitative evolution of a particular operation, suitably isolated from temporally and geographically distant or intuitively unrelated events."[84] Pure mathematicians hailed axiomatics for its universalizing possibilities. Axiomatic definition abstracted away from particular contexts and thus offered a more generalized theory that could be applied in a myriad of cases. This same impetus now drove applied mathematics. A model was deemed successful exactly because it was able to isolate the essential features of a phenomenon without being hampered by "intuitively unrelated events."[85] The structure of the model, not the context of its emergence, was what drove the analysis.

The use of mathematics for measurement, quantification, and calculation did not diminish after the war, but the report's commitment to the axiomatic conception of mathematics entailed that all these uses of mathematics were presented as secondary to its "true" use in theoretical construction. The report did not discuss how the calculus, partial differential equations, or functional analysis could be applied to physical, biological, and social phenomena, but rather sought to identify the "mathematical modes of thought" that defined contemporary application of mathematics. In case after case, the report considered the nature of mathematical formalism. The author of the report was Joachim Weyl, who immigrated with his father, Hermann Weyl, to the United States in 1933. Following in his father's footsteps, the young Weyl received a doctorate in mathematics from Princeton University in 1939. His research was originally in pure mathematics, but during the war his interests shifted as he began working for the Office of Naval Research (ONR). By 1949, he was appointed chair of the Mathematics Branch at the ONR. He had close and intimate knowledge of *all* the emerging sciences of war and he conceptualized all of them in terms of axiomatic analysis.

The same was the case with mathematician Richard Burington, who served as the head mathematician in the Bureau of Ordnance in the Navy. In 1949, he published "On the Nature of Applied Mathematics," in which he exhaustively analyzed mathematization as axiomatization. Regardless of the underlying situation, the job of the mathematician, in Burington's account, was model construction, which was necessary in order to strip away "the lesser distracting elements, leaving only the essential and fundamental features, skeleton and flesh."[86] This act of model construction was remarkably similar to the process Veblen and Birkhoff had described a decade earlier. In particular, Burington explained that when well-established concepts and intuitive notions do not work, it might be "necessary to use a high type of free mathematical construction, using premises which may appear to have little connection with previous experience as a guide."[87] As long as the theory's deduction could be subjugated to experience, it was irrelevant how abstract the premises might be. Burington acknowledged that his view of model construction built directly on the "postulational and deductive character of pure mathematics." Indeed, he even offered his reader a crash course in axiomatics.[88]

However, there is a difference between his analysis and that of his predecessors. When Birkhoff and Veblen were writing, they had theoretical physics in mind. Axiomatization was a means of uncovering the laws of nature. Burington had something else in mind. Model construction was motivated by practical concerns. The models were idealizations whose goal was finding solutions to problems. Burington wrote, "No matter how logical and beautiful an analysis is made in connection with an industrial project of consideration, if it fails to meet the realist tests of relevance, adequacy, and applicability, sooner or later the work will be recognized . . . as being inadequate."[89] This was the greatest difference between interwar and postwar mathematization. The meaning of mathematization remained remarkably similar, but the purpose had changed. After the war, applied mathematicians found themselves in new institutional settings answering to new lords, but the intellectual justification for their approach was rooted in the interwar period.

By the 1950s, there was almost a complete consensus among mathematicians that the job of the applied mathematician was *formulation*. Griffith E. Evans, who chaired the Department of Mathematics at the University of California, Berkeley, is credited with transforming the department into a top-tier research center. During remarks on the nature of applied mathematics in 1953, he explained, "With reference to the objective universe as it is portrayed in the physical sciences, the life sciences and the social sciences, the *mathematical task* is that of employing the method of abstraction to the point of maximum concrete return . . . It is the method of generalization of a particular problem into a systematic theory."[90] Particular problems were important, Evans wrote, but "what the mathematician [could] offer beyond special techniques [was] a general approach, a comprehensive grasp of symbolic structure, a systematic survey."[91] Evans's remark perfectly captured mathematicians' conception of mathematical analysis at midcentury. Adopting the lessons they learned in pure mathematics, mathematicians brought to the study of the world (physical, biological, and social) their emphasis on abstraction, generalization, and structure—all markers of high modernist mathematics.

Other mathematicians similarly insisted that the mathematician's role was to provide theoretical understanding. In reaction to Evans, Abraham Taub, a mathematician and physicist, noted, "Few mathematicians realize as yet that half of the applied mathematician's task is that of finding and formulating mathematical problems—once this is done any mathematicians can work on them."[92] Indeed, while applied mathematicians were unable to settle on a definition of applied mathematics, they all seemed to agree that it was *formulation* that was the fundamental role of the applied mathematician, not calculation.[93] Defining the basic concepts of a theory and the relations between them was a fundamentally mathematical act. Once this initial step was accomplished, the problem could be solved by any mathematically inclined individual, but it was not and should not be confused with the true task of the mathematician.

John Tukey fought most forcefully to establish this new vision of mathematization. He is emblematic of the generation of mathematicians who turned toward applications following their experience during World War II. Tukey arrived at Princeton in 1937. He was originally planning to pursue a PhD in chemistry, but was ensorcelled by the siren song of mathematics. He earned his degree in topology just two years later (in the same class as Norman Steenrod). Lined up on the graduation stage beside him was none other than Joachim Weyl. During the war, however, Tukey's research interests began to shift. In 1941 he joined the Fire Control Research Office at Princeton, which was headed by mathematician Merrill M. Flood. While he had no interest in statistics before the war, by 1945 he had begun to self-identify as a statistician, and he soon took a joint appointment, splitting his time between the Department of Mathematics at Princeton and Bell Laboratories. His experience made him sensitive to the plight of applied mathematicians. It had been at Tukey's urging that Synge's committee gathered in 1946 to study the place of applied mathematics in the AMS. Earlier that year, Tukey had sent a letter to mathematicians around the country announcing a "crisis in applied mathematics." The society began to study what it could do to promote applied mathematical research in the United States in response to Tukey's agitation.

Tukey's interest in applied mathematics did not falter in the following decade, and he continued to campaign on its behalf. During his career, Tukey had carved out for himself a unique advisory position across multiple subfields: he worked on antiaircraft missile systems and nuclear weapons, served as a member of the National Security Agency Scientific Advisory Board, consulted for the government on the census, and was entrusted by the American Statistical Association to review Alfred Kinsey's report *Sexual Behavior in the Human Male*. His own experience convinced him that the true role of the mathematician was as a type of consultant. More than knowledge of a specific theory or technique, mathematics was a way of thinking.

In 1958, Tukey presented some of his ideas in "The Teaching of Concrete Mathematics," in which he elucidated his vision for the field. One of the most promising areas for applied mathematics, he insisted, was the development of a theory of formulation. "It is agreed that formulation of the problem is usually the most important stage in 'applied mathematics,'" he began.[94] Formulating new concepts and refining old ones was essential to this endeavor. Echoing the words of Birkhoff before him, Tukey mused, "Insofar as a concept-former is a philosopher, all mathematicians need to be philosophers (of a very special sort)."[95] Yet even though formulation of a problem was the essential task of the mathematician, no one had ever explicitly studied this problem of formulation. "Who has tried to explain it to the student? Polya wrote 'How to Solve It'; who will now write 'How to Formulate It'?"[96] This, Tukey insisted, was the most important task facing applied mathematicians, for until they knew how to teach students the art of formulation, they would be unable to produce applied mathematicians. Tukey held that the new vision of mathematization required a new kind of mathematician.

The impact of this new conception of mathematization as axiomatization and formulation was far-reaching. On the one hand, as I show in chapter 3, not only mathematicians but social scientists as well began advocating for this new and expanded image of mathematization. Researchers in the human sciences debated and struggled most forcefully with questions about the meaning of theoretical knowledge and the role of mathematics in explicating social phenomena. However, as will become clear in chapter 5, this broad vision of mathematization also held back the mathematical community. Despite the hard work of researchers like Tukey, a new type of applied mathematician did not emerge in the following decades.

CHAPTER THREE

Human Abstraction

"The Mathematics of Man" and Midcentury Social Sciences

It has always been the curse of philosophy (until this curse was lifted by the logical positivists) to assume that entities called politics, society, power, welfare, tyranny, democracy, milieu, progress, etc., actually exist just as cats, icebergs, coffee pots, and grains of wheat exist, and that each had an essence discoverable by proper application of reason and observation.
—Anatol Rapoport

In 1954, Claude Lévi-Strauss announced a "revolution" in the mathematization of the social sciences. For Lévi-Strauss, the work of the preceding decade had demonstrated that a new "mathematics of man" was emerging, one that broke firmly with past attempts to utilize mathematics in social scientific research. "Social scientists were certainly aware, well before the last ten years," Lévi-Strauss explained, "that a science is not really a science until it can formulate a precise chain of propositions, and that mathematics is the best means of expression for achieving this result."[1] In psychology, economics, and demography, Lévi-Strauss said, researchers had appealed to mathematical techniques to advance their studies, but their efforts to date had been limited. This was not because, as some held, social phenomena were less amenable to mathematics than natural phenomena, but because those researchers misunderstood the nature of mathematics. Lévi-Strauss wrote, "[They restrict themselves to] quantitative methods which, even in mathematics itself, are regarded as traditional and largely outmoded, and have not realized that a new school of mathematics is coming into being and is indeed expanding enormously at the present time—a school of what might almost be called qualitative mathematics, paradoxical as the term may seem, because a rigorous treatment no longer necessarily means recourse to measurement."[2] Mathematics, Lévi-Strauss insisted, did not equal quantification.

As Lévi-Strauss explained, he himself came to this realization after collaborating with Bourbaki member André Weil on a "rigorous treatment" of the rules of marriage and descent.[3] "The mathematician has absolutely no need to reduce marriage to quantitative terms; in fact, he did not even need to know what marriage was," he explained to his readers.[4] Lévi-Strauss is most associated with the school of French structuralism, but he understood the mathematization of the social sciences to be a broader phenomenon. As examples of the new "mathematics of man," Lévi-Strauss pointed both to game theory and a recent book by sociologist Paul Lazarsfeld (more on that soon). Lévi-Strauss emphasized that the fruitful collaboration between mathematicians and social scientists would undoubtedly take advantage of new schools of thought, such as set theory, group theory, and topology, all of which emphasized relations between classes. Hence, in adopting mathematics, social scientists needed to move beyond quantification. This new "mathematics of man," he explained, "is resolutely determined to break away from the hopelessness of the 'great numbers'—the raft to which the social sciences, lost in an ocean of figures, have been helplessly clinging."[5]

Lévi-Strauss's prognosis that what would define the mathematization of the social sciences in the post–World War II era was the *de*identification of mathematics with quantification was shared by many social scientists during the period. Computer scientist and mathematician John G. Kemeny perfectly captured this sentiment in an article he wrote on the social sciences titled "Mathematics without Numbers."[6] As will become clear in this chapter, while the psychologists, sociologists, and economists who used mathematics in their work differed in their research agendas, they shared the belief that mathematization did not equal quantification. Rather, like the mathematicians described in chapter 2, they believed that mathematics' greatest promise was as a tool for theory construction. Mathematics, they held, offered conceptual clarity. They called for the adoption not of a particular mathematical technique—although several methodologies were increasing in popularity—but for *mathematical thinking*.

Historians of science have so far followed the work of Ted Porter in trying to explain the mathematization of the human sciences in the postwar period. In this view, mathematics is equated with trust in numbers and objectivity.[7] However, when postwar social scientists appealed to mathematics, it was *not* quantification they sought, as Lévi-Strauss insisted. What they were after, as I will show, was *axiomatization*. Other scholarship on the postwar social sciences amply describes Norbert Wiener's cybernetic theory and elucidates its influence on social scientific theory.[8] While Wiener's cybernetic theory is not an example of axiomatic thinking per se, Wiener's earliest publications belonged to the uniquely American school of postulational analysis, which emerged as a response to Hilbert's axiomatics. At Harvard, Wiener was influenced by E. V. Huntington, who was one of the group's most prolific authors. That he was steeped in modern axiomatics is obvious. In emphasizing the influence of axiomatics, my goal is different from histories of cybernetics. Rather than showing how certain concepts, analogies, or ideas about human nature trafficked freely among a diverse set of postwar scholars, I show how axiomatics altered what counted as *theory* in the postwar social sciences tout court. Namely, while historians have correlated mathematics with positivism and objectivity in the postwar decades, its most sweeping effect was rather in redefining the meaning of *theory* on axiomatic grounds. Indeed, the "mathematization" of the social sciences was felt even in studies that used little in the way of mathematics. Axiomatics reshaped what theorizing entailed.

Historians of the social sciences have also noted that the postwar period marked a shift in social scientific research in the United States.[9] During the interwar period, as Joel Isaac notes, social scientists "placed their faith in measurement and instrumentation, treating the accumulation of statistics and observational data as the cardinal scientific virtues."[10] Only in the aftermath of World War II did theory construction become a dominant feature of social scientific work. Isaac traces this transformation to the 1930s: the impact of relativity theory and quantum mechanics, the growing popularity of conceptual frameworks such as "system" and "organism," the influx of European émigrés, and the mathematical formalization of economics all contributed to the belief "that theoretical abstraction had a role to play in the discovery of social laws."[11] What form did these theoretical abstractions take? To what degree were they modeled on axiomatics? And what was the relation between mathematization and the meaning of theory in the social sciences?

I follow the writings of a host of American social scientists in the postwar period, including Howard Raiffa, Anthony Downs, and Anatol Rapoport, in order to unpack how *they* understood the appeal of mathematics during the period. Why was mathematics useful, or even necessary? From economics to psychology, sociology, and political science, *axiomatic reasoning* offered a shared approach and a shared set of commitments. In the hands of mathematicians, axiomatics designated a well-defined procedure, with an emphasis on the consistency and independence of axioms; but its appropriation into the social sciences did not follow such strict logical constraints. What, then, defined axiomatization in the human sciences? First and foremost, as in mathematics, modern axiomatics designated a break between analytic formulation and post hoc empirical verification. This, perhaps, was its strongest legacy.

The earliest example of this change can be traced to 1940, when psychologist Clark L. Hull from Yale's Institute of Human Relations published *Mathematico-Deductive Theory of Rote Learning*.[12] The book, written in collaboration with three mathematicians, offered a complete postulational theory of how people acquire knowledge through repetition.[13] Hull set up a system with sixteen undefined terms, eighty-six definitions, and eighteen postulates, which he then used to deduce theorems that could be verified via observation. Hull and his collaborators believed the book, more than just a contribution to learning theory, offered an example of a scientific study of human behavior. In the introductory essay, Hull explains that while it might be necessary at the start of a study to take an informal approach, a formal treatment can help "facilitate the removal of presumptive fallacies."[14]

Reviewing the work for *American Anthropologist*, Gregory Bateson wrote that the book's main contribution to anthropology was that "it provides us with *a lesson in how to think*, and even those who do not propose to work through the massive apparatus of definitions, postulates and theoreums [*sic*] will do well to read the thirteen pages of introduction, *Concerning Scientific Methodology*." Bateson added, "The naive will discover that when knowledge is formally exposed, the order of procedure is the reverse of that adopted in reporting upon research. The ordinary scientist thinks inductively, deriving his general conclusion from his particular observations, but in this book the most abstract conclusions become 'Postulates' from which the author *de*duces the phenomena which can be observed or ought to be observable in the laboratory."[15] More than a decade before Lévi-Strauss announced the mathematics of man, his American colleague equated mathematics with a new way of thinking.

While historians have noted that the postwar transformation of the social sciences was characterized by an increased adoption of mathematics, mathematics tout court is an insufficient explanation.[16] Further, the degree to which high modernist social sciences embraced axiomatic reasoning remains unacknowledged.[17] Hunter Heyck argues that, beginning in the 1950s, "the subjects of social science were systems structured by relations, the method employed was behavioral-functional analysis and the goal was a theoretical model, one that potentially could be made an operational guide to practical action—in some other, future publication."[18] Heyck emphasizes that this new approach to the social sciences relied heavily on the adoption of mathematics, but for him it was social scientists' propensity to "see the world as a system" that "encouraged mathematical formalization."[19] Considering the number of mathematically trained individuals whose work was fundamental to the transformation Heyck describes, the arrow of causality might be reversed, or at least point in both directions. The adoption of axiomatic thinking did not in itself usher in the belief that social science theories were "systems structured by relations," but it certainly played an important role.[20]

Moreover, an attention to axiomatics illuminates what distinguished the positivistic nature of postwar social sciences. In *The Politics of Method in the Human Sciences*, George Steinmetz emphasizes the influence of logical empiricists and logical positivists on the conception of scientific knowledge in human sciences during the postwar period. Positivism, which as Steinmetz acknowledges in the introduction, dates all the way to Hume, is defined as the idea that theories must take the form of "if . . . then."[21] An understanding of theoretical knowledge as a collection of postulates from which theorems can be derived was prevalent among human scientists. However, what distinguished the postwar human sciences was an uneasiness with the way this analytic construct was connected to the "real" world, in particular a willingness to except axioms as unverifiable premises.[22] When applied to the social sciences, the main influence of axiomatics was in creating a break between analytic formalism and empirical content. Axioms no longer needed to be derived inductively or taken to be self-evident truths. As with mathematics, theory in postwar human science became abstract when it was no longer believed to be abstracted.

I start by looking at social scientists' direct efforts to bring mathematics into their field. As I show, social scientists across several research fronts appealed to axiomatics as among the most promising tool for mathematization. They elevated abstract or mathematical thinking above any particular mathematical technique. I then interrogate what social scientists considered to be the utility of axiomatics. As I demonstrate, they followed pure mathematicians' emphasis on (1) economy of thought, (2) structural relations, and (3) the belief that any true essence is hidden beneath the surface. Finally, I examine rational choice theory, the most well-known example of axiomatic thinking in the social sciences. Its use here illuminates the defining feature of postwar axiomatics, namely the separation of analytic formalism from empirical phenomena. However, as I show, this bifurcated view of theory forced social scientists to reevaluate the very meaning of theory in the social sciences more broadly.

Bringing Mathematics In

Lévi-Strauss's "The Mathematics of Man" served as the introduction to a special issue of the *International Social Science Bulletin*, which grew out of a "Seminar on the Use of Mathematics in the Human and Social Sciences." The seminar had taken place at UNESCO House in 1954, under the auspices of the International Social Science Council. The issue includes articles on the use of mathematics in sociology, economics, and public opinion, as well as reports on research centers where such work was being conducted. The goal was to encourage the use of mathematics in the social sciences by providing numerous illustrations of its application. Such efforts were common on the other side of the Atlantic. Between 1953 and 1957, the Social Science Research Council (SSRC) sponsored five summer institutes in mathematics for PhD students, postdoctoral fellows, and professors seeking to improve their mathematical literacy. They were organized by an SSRC Committee on Mathematical Training of Social Scientists, whose job was to promote such endeavors. Chaired by mathematical statistician William Madow, membership of the committee over five years included psychologist Robert R. Bush, psychologist W. K. Estes, sociologist E. P. Hutchinson, computer scientist John G. Kemeny, economist Jacob Marschak, cognitive psychologist George A. Miller, system analyst Howard Raiffa, economist Robert Solow, and operational researcher Robert M. Thrall.

What is telling is not that the SSRC decided to support mathematical methods in the social sciences, but rather the type of mathematics that was deemed appropriate for budding researchers. In 1953, when the first summer institute took place at Dartmouth College, the instructors for the course, including Bush, Thrall, Raiffa, and Madow, identified four areas of research: probability, calculus, algebra, and axiomatics.[23] The inclusion of axiomatics, in particular, points to the mathematization of theoretical knowledge in the postwar period.[24] Thrall, who was trained in pure mathematics, was among many young mathematicians who redirected their research from pure mathematics to operations research and allied fields during World War II. He was a member of the mathematics department at the University of Michigan. Reflecting on his experience at the 1953 summer institute, Thrall explained that, especially when it came to model building, "a considerably greater emphasis on axiomatics and foundations" was necessary.[25] Raiffa similarly believed that axiomatics had an important role to play in social scientific investigation. However, he emphasized that it was not a complete axiomatic treatment, such as that given by von Neumann and Morgenstern, that would be useful. It was, rather, "*axiomatic-type thinking*" that was necessary to bring clarity to otherwise "nebulous material (e.g., cohesiveness of a group, so-called 'rational' behavior, 'socially desirable' welfare functions, etc.), definition after definition is discarded or discredited because it does not fit the bill in certain situations."[26] To avoid such situations, Raiffa recommended, researchers should state clearly in terms of axioms those conditions they wished their definitions to satisfy, and "then determine whether they [were] consistent, independent, and categorical."[27] The axiomatic definition of concepts was, after all, a hallmark of the modern axiomatic approach.

Raiffa emphasized that the reason to include axiomatics was not because it offered a more sophisticated approach to the social sciences, but rather because it directed the researcher's "attention to the structure and to some of the basic notions of the area." He added, "Axiomatic type thinking has played and will play in the future an increasingly important role in mathematical work in the social sciences."[28] What Raiffa wanted to highlight was the "importance of abstract thought (in contradiction to mathematical technique) to the social scientist."[29] Axiomatic-type thinking had the potential to accomplish just that. It was not so much about developing strict axiomatic systems as it was about forcing researchers to clarify their assumptions and their potential ramifications. Even Madow, who noted that a controversy existed between those who believed axiomatic treatments were to be promoted and those who opposed them, concluded by noting that "considerable gains" would come to social scientists "from the consciousness that [came] from the axiomatic approach, of the way in which mathematics and the world of experience interact."[30]

In addition to the summer institutes, the SSRC also supported the publication of textbooks that would help social scientists obtain the necessary mathematical skills. In 1956, Bush, who was an instructor at the first summer institute, published *Mathematics for Psychologists: Examples and*

Problems, together with Robert Abelson and Ray Hyman.[31] Bush was one of many physicists who turned away from physics in the aftermath of World War II. While still at Princeton, where he earned his PhD, Bush organized a monthly seminar with fellow students, asking: "Are the methods of physical sciences applicable to social sciences?" In 1949, he took a postdoctoral position at Harvard's Department and Laboratory of Social Relations, where he turned his attention to psychological research. After his postdoctoral appointment finished, he remained in the department, where he taught, among other courses, a seminar in mathematical models in the social sciences. In this seminar, he drew upon examples across multiple fields (including game theory, models of learning, kinship structures, and growth models).

In *Mathematics for Psychologists*, Bush once again identifies four areas of mathematics he deemed relevant for psychologists: calculus, probability, matrix algebra, and *mathematical foundations*. There is no obvious reason psychologists should study the foundations of mathematics or learn the language of set theory unless it is recognized that such training serves as an introduction to axiomatic reasoning. In the introduction to the book, the authors single out four mathematical textbooks that parallel the book's organization and can serve as guides for the mathematical material. They recommend *The Anatomy of Mathematics*, published in 1950, to readers who want to learn more about mathematical foundations. Written by mathematician turned engineer R. B. Kershner and mathematician L. R. Wilcox, *Anatomy* was an ode to axiomatics. In the preface, the two authors explain that they conceived of the book as a "treatise on the axiomatic method." Their goal was to offer an introduction to modern mathematics that did not assume any prerequisite knowledge on the part of the reader. Emphasizing the point, the authors write that not only does the reader not "need to know the sum of 7 and 5," but "he will not learn it from this book."[32] This, in part, was the appeal of axiomatics.

Kershner and Wilcox's commitment to presenting mathematics through axiomatics is illustrated in the first substantive chapter, which is not about numbers or even set theory but instead comprises an analysis of "language." The authors begin by discussing the difference between spoken and symbolic language and the potential for confusion that might arise by the use of either. Using the statement "She was fair" as an example, the authors explain that if you are unclear about the meaning of the word "fair," turning to the *Oxford English Dictionary* won't solve your problem. Does the statement imply that the "she" in question was "beautiful, clean, impartial, or just otherwise admirable"?[33] Such multiplicity of meanings might be palatable in ordinary language, but in a situation in which a statement should have meaning outside of context, terms "*must be unambiguous*." But the confusions of language are even greater: Many definitions are cyclical, and while the meaning of a word like "horse" can be learned relatively easily in childhood, more abstract terms like "beauty" are harder to define. An elaborate discussion of the ambiguities of language at the opening of what is essentially a mathematical textbook is a way of acclimating the reader to axiomatic reasoning. As the authors explain, some words must remain undefined ("There should then be a basic list of words that we forgo defining."). *The Anatomy of Mathematics* appeared repeatedly on social scientists' reading lists in the following decade.

The efforts of SSRC were replicated on a smaller scale in other institutions. Starting in 1950, a seminar on the application of mathematics to the social sciences took place at the University of Michigan. The seminar was sponsored by the departments of economics, mathematics, philosophy, psychology, and sociology.[34] The success of the seminar's first two years encouraged the organizers to apply for a grant from the Ford Foundation for an eight-week summer seminar, "The Design of Experiments in Decision Processes," to be held in Santa Monica, California. The venue for the event was chosen, not surprisingly, for its proximity to RAND, and the participants included mathematicians, philosophers, psychologists, and economists. In the proceedings of the conference, *Decision Processes*, the editors (including Thrall) explained that the goal of the gathering was to coordinate and find common ground among the "rapid development in the mathematical formulation and testing of behavioral science theory."[35] The editors then located the "modern" conception of decision problems in two relatively recent mathematical developments: von Neumann's formulation of utility in game theory and Abraham Wald's statistical decision theory. As they explained, decision problems—namely, deciding whether a given action would be apt for achieving a desired effect—had occupied philosophers for more than two thousand years. What was new was the mathematical "framework," or "abstract system," that researchers applied to the problem.

In *The World the Game Theorists Made*, Paul Erickson follows the work of these mathematicians, demonstrating how important their research was in the early development of game theory in the human sciences.[36] As Erickson explains, knowledge of game theory, especially in the early days, passed between individual mathematicians via direct interactions. Erickson also makes clear that the diffusion of mathematics at the time was enabled by a set of institutions and financial networks, such as the Ford Foundation and Stanford's Center for Advanced Study in the Behavioral Sciences. For Erickson, the proliferation of game theory, or what he terms the "game theory phenomenon," can best be explained by approaching game theory not as a coherent entity but as a collection of what historians of science term "theoretical tools." Beyond these theoretical tools, which they dispersed across research centers, mathematicians' greatest impact, I would argue, was in insisting on the separation between analytic formalism and empirical phenomena. This bifurcated view of theory was fundamental to axiomatics and became, in the postwar period, the main characteristic of mathematization.

In the opening paper, "Mathematical Models and Measurement Theory," Raiffa, Thrall, and psychologist Clyde H. Coombs defined in detail mathematical models and their use in measurement theory. Not surprisingly, the authors began by offering a quick explanation of axiomatic reasoning. They also directed the interested reader, who might wish to learn more about the nature of mathematical systems, to three prior works: *The Anatomy of Mathematics*, Weyl's *Philosophy of Mathematics and Natural Science*, and Wilder's *Introduction to the Foundations of Mathematics*.[37] All works emphasized modern axiomatics, and all were written by mathematicians, not logical positivists. The translation between "real world" and model, the authors explained, was one of abstraction and interpretation, or abstraction and realization (see Fig. 3.1). Expanding on this diagram further, the authors explained that, in this view, the task of science is to arrive at the same conclusion by two paths: experimentation and logical deduction. Throughout the paper, the authors sought to account for a movement between mathematical formulation and the experimental situation, the a priori and its realization. "The model becomes a theory about the real world" only after a "segment of the real world has been mapped into it."[38] A model can only be evaluated on logical grounds; a theory must satisfy "external criteria." Thus, while according to the diagram the model is arrived at by abstraction from the world, this by itself does not make it a model *of* the world. At this stage it can still be judged only on "logical grounds." The model becomes a theory of the world only after the elements of the model are given real world interpretation. This bifurcation was characteristic of axiomatics. Further, the axiomatization of measurement theory, which the three editors developed in their paper, makes clear that, by midcentury, measurement and quantification were subordinate to axiomatics.

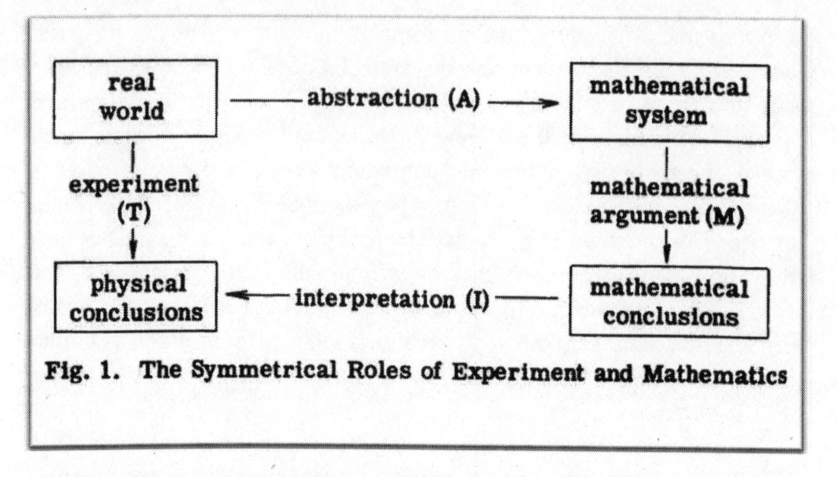

FIGURE 3.1 A diagram from "Mathematical Models and Measurement Theory," illustrating the relations between a mathematical model and the real world in terms of abstraction and interpretation.

Source: *Decision Processes*, ed. R. M. Thrall, C. H. Coombs, and R. L. Davis

Raiffa's path to mathematics and statistics was meandering. At the age of 22, after completing his military service, he went to the University of Michigan to study actuarial mathematics. During his time there he took a course in the foundation of mathematics taught in the R. L. Moore tradition.[39] Like Steenrod more than a decade earlier (see chapter 1), the experience transformed Raiffa, and he decided to pursue a PhD in mathematics. His interest in operations research and game theory grew while serving as a research assistant on an Office of Naval Research (ONR) project sponsored by the Department of Mathematics and the School of Engineering. A research report he wrote turned into his dissertation, and after a year as a postdoc at Michigan, during which he helped edit *Decision Processes*, he found himself teaching statistics at Columbia University.

At Columbia, Raiffa started working with R. Duncan Luce, who had been hired to supervise an interdisciplinary project on behavioral modeling. Luce received his PhD in mathematics from MIT, working on abstract algebra, but he soon applied his mathematical knowledge to psychological questions. Raiffa and Luce's collaboration resulted in the publication of *Games and Decisions*, a work that has been described as "the most influential event in the literature of the period for the spread of game-theoretic thinking by the early 1960s."[40] More than von Neumann and Morgenstern's original, Luce and Raiffa's book, *Games and Decisions*, helped propel game theory into the social sciences. This was at least in part due to the authors' decision to reduce the mathematical content to a bare minimum. In the introduction, Raiffa and Luce explain that while knowledge of calculus and matrix algebra would be useful, it is not necessary. What is necessary is "mathematical sophistication," which, the authors explain, amounts to the reader's willingness to "accept conditional statements, even though he feels the suppositions to be false."[41] The reader must be willing "to make concessions to mathematical simplicity" and, above all, "must have sympathy with the method."[42] Such a definition of mathematical sophistication presupposes that, by this point, mathematics was equated with a particular way of reasoning, or an approach, rather than a calculative technique.

Columbia University, where Raiffa and Luce's collaboration took place, was also teeming with mathematical activity. Paul Lazarsfeld organized a seminar on mathematical sociology together with Ernest Nagel, and a seminar on mathematics in the social sciences convened in the evenings. Lazarsfeld, whose work had been foundational to the growth of American sociology during the period, invited researchers whose work he found interesting. James Coleman, who was a student at Columbia at the time, remarked that, to a certain degree, one could characterize mathematical sociology at Columbia during the 1950s as a "collection of people he [Lazarsfeld] had gathered around himself for the purpose of teaching him[self]."[43] In 1954, Lazarsfeld collected several of the lectures presented at Columbia and published them in a new edited volume titled *Mathematical Thinking in the Social Sciences*. The title is revealing: It was *mathematical thinking*, rather than any particular mathematical technique, that Lazarsfeld believed had the most to contribute to the social sciences.[44] Lazarsfeld acknowledged that the use of mathematics in the social sciences had been controversial; he suggested, however, "Even those who do not believe in any early use of mathematics try to utilize some rudiments of formalization in order to clarify their underlying assumptions, and to derive specific findings from more general models."[45] Mathematics, in his account, provided a way to organize knowledge.

Angela M. O'Rand identifies three strands in the mathematization of the social sciences in the 1950s. In addition to game theory, she writes, the two other main research fronts were the work of Kurt Lewin at MIT and that done under the direction of Paul Lazarsfeld at Columbia University's Bureau of Applied Social Research. O'Rand writes that "the technical core of each of these programs was not the mathematical proofs as in the case of game theory, but the social experiment and the social survey, respectively."[46] Despite these differences, social scientists from different fields agreed for the most part on the nature of mathematization. Namely, they might appeal to different mathematical methods or put emphasis on different aspects of a research program, but in a general sense they agreed on why mathematics was necessary in theoretical work. They all held that mathematics offered clarity of thinking, not computational capacity. Conceptual clarity and an emphasis on a relational approach were at the core of mathematization, and hence of theorizing.

In his introductory essay to the edited volume, Lazarsfeld praised SSRC's effort to promote instruction in mathematics to social scientists. He did not necessarily think mathematization was the only path forward, but he identified it as one with great potential benefits. With the edited volume, his goal was to "take specific problems and look at them with the end in mind of understanding better how the structure of behavioral science thinking and the structure of various mathematical methods fit each other." An emphasis on structure, of course, was a central feature of axiomatic thinking, and Lazarsfeld addressed the "flowering of axiomatics," suggesting that it could offer clarity in the social sciences as well: "Is it not possible that a similar type of analysis would help to explicate concepts of the social scientist and throw some light on what alternatives he might have overlooked?"[47] Lazarsfeld's work at the Bureau of Applied Social Research was known for its empiricist tendencies. The bureau was recognized for the development of survey analysis and the technique of latent structure, not for grand theorizing. But even Lazarsfeld acknowledged that mathematics' potential in the social sciences was not restricted to quantification.

Lewin's work stood out because he appealed to topology rather than calculus, statistics, or differential equations in his theories. Lewin was drawn to topology after reading Veblen's *Analysis Situs*. He sought to integrate the structural conception of modern mathematics in his theories and, like Lévi-Strauss, he believed that modern mathematics was necessary because past efforts to collect and measure had been limited. In *Formalization and Progress in Psychology*, published in 1940, Lewin reflected on the limited impact of psychological research that focused on strict observation and measurement. "After all," he wrote, "what psychologists observed were human beings. Children needed help and education; delinquent people needed guidance; people in distress wanted cure. Counting, measuring, and classifying their sorrow did not help matters much." He then added, "Obviously one had to go to the facts 'behind,' 'below the surface.'"[48] The distinction Lewin draws—between understanding a given social phenomenon by counting and measuring what is on the surface, and discovering its hidden structure—captures the shifting conception of mathematization between the interwar and postwar period.

Needless to say, many social scientists continued to appeal to mathematics for its computational capacity. Far from diminishing in popularity, the use of statistics in psychology and sociology only increased during the postwar period. The collection and analysis of experimental results became mired in numbers as researchers learned how to conceptualize uncertainty and confidence intervals.[49] Still, to the midcentury scholars described in this chapter, this prevalent use of computational mathematics was secondary to the role of mathematics in theoretical formalization. For them, mathematics offered conceptual clarity and drew researchers' attention to structural relations. Moreover, by midcentury social scientists were praising axiomatics for its economy of thought and, as was already evident in Lewin's writing, its ability to peer beneath the visible surface.

Why Axiomatics?

Three years before Lazarsfeld's edited volume appeared, Hermann Weyl surveyed the previous half-century of mathematics in the *American Mathematical Monthly*. The article focused on axiomatics, which Weyl identified as "one very conspicuous aspect of twentieth century mathematics."[50] Weyl, who was suspicious of the increased emphasis on axiomatization and abstraction in pure mathematics, nonetheless identified some of the strengths of the method. First, axiomatics offered clarity and simplification. "The real aim is simplicity," he explained. When faced with a complex mathematical situation, he wrote, the real art was to be able to separate various sides of a subject and generalize from them. Such generalizations would simplify as they reduced the number of assumptions taken into account. Thus, Weyl wrote, "The basic notions and facts of which we spoke are changed into undefined terms and into axioms involving them."[51] But beyond simplifying things, axiomatics promised a certain economy of thought: when an axiomatic system had been established and a body of statements had been deduced from it, they were then available indefinitely, "not only for the instance from which the notions and axioms were abstracted, but wherever . . . an interpretation of the basic terms which turns the axioms into true statements" was encountered.[52] This ability to reinterpret the same axiomatic system over and over again was taken as a general benefit of axiomatics.[53]

Weyl had pure mathematical research in mind, but the belief that axiomatics offered economy of thought migrated into the social sciences as well. One of the contributors to Lazarsfeld's edited volume was Nicolas Rashevsky, an expert in mathematical biology and biophysics who had an interest in social behavior. Rashevsky's contribution to the volume offered a model for the behavior of individuals under the influence of environmental conditions. The paper, however, was quite technical, and Lazarsfeld was not pleased with the written result. He asked one of Columbia's doctoral students in sociology, James Coleman, to write an addendum chapter offering an expository analysis of Rashevsky's work that would be more accessible to the reader. Working through Rashevsky's theories, Coleman offers a lesson in model building, emphasizing the ability to reinterpret a model in numerous situations. "The premise which is implicit in our whole approach," writes Coleman, is the independence of "the two parts of any mathematical model: the *structure*, which is completely mathematical, devoid of any social or physical meaning, and the *content*, which consists of the meaning given to the various parts of the structure."[54] Coleman goes on to show how the same structure can apply to different contents. For example, one could take a model that examined the distribution of *attitude* in individuals and change it to a model studying the distribution of *status*, or one analyzing the *conformity* of individuals to one describing their *satisfaction*. All that a researcher would need to do in order to reinterpret the model would be to assign a different meaning for the variables in the system. In the words of Raiffa, Thrall, and Coombs, one would only have to "map" a different "segment of the world" onto the model.

A still more radical transformation was possible. All of Rashevsky's models, Coleman explains, discuss social behavior, but they can also be used to study economic activity. He writes, "If we wanted to use one of them, say the model of 'altruistic and egotistic societies,' for representing the behavior of competing firms in economic systems, we would let the elements be firms rather than individuals."[55] In other words, the clear separation between the analytic model and observation—the "lonely existence," to borrow Veblen's words, of the model—implied that it could be given different interpretations and hence apply to different social circumstances. In his introductory essay, Lazarsfeld notes that this aspect of Coleman's exposition "might in the long run prove most helpful," as it stresses the relation between the mathematical elements of the model and concrete social situations.[56] Regretting that not enough comparative literature from the social sciences exists, Lazarsfeld directs the reader to *The Anatomy of Mathematics*.

Lazarsfeld was not the only researcher who believed this capacity of formal analysis was worthy of high praise. In the 1950s, Lewin's emphasis on topology was taken up by other mathematicians and psychologists. In 1953, mathematicians Frank Harary and Robert Z. Norman, working at the Research Center for Group Dynamics at the University of Michigan, an offshoot of Lewin's MIT center, published "Graph Theory as a Mathematical Model in Social Science," urging others to adopt this method in psychological and sociological research. "Why should a social scientist be interested in separating the formal aspects of the subject from its concrete sociological or psychological setting?" the two ask at the opening of their paper. The answer, at least in part, was that the same mathematical model ("a set of unproved statements called postulates or axioms, a set of undefined terms called primitives, and the collection of all theorems deducible from these postulates and the laws of logic") could be used in multiple situations. "The power of this abstract approach," they write, "is that the theorems of a mathematical model give information about each and every interpretation, i.e., concrete systems or realization, satisfying the postulates."[57] The results, they explain, are obtained by "a simple translation" of the theorems.[58]

Economy of thought, however, was only one of the strengths of axiomatic reasoning that Weyl identified in his 1951 essay. Another impetus behind axiomatic analysis was that it often helped reveal "inner relations between . . . domains that apparently lie far apart." Leaving the realm of pure mathematics to make his point, Weyl writes that "it is as if you took a man out of a milieu in which he had lived not because it fitted him but from ingrained habits and prejudices, and them allowed him, after thus setting him free, to form associations in better accordance with his true inner nature."[59] The true nature of this hypothetical man, Weyl explains, can be revealed *only* once the man is "free" from his environment. In other words, axiomatics was powerful because it offered a tool with which to identify the essence of a problem unencumbered by superfluous details. This message, too, was adopted by social scientists. Lazarsfeld explained, for example, that formalization in social science, even if simplistic, was useful when it could "highlight the *essential* features of a substantive problem."[60] Historians have often remarked on the prominence of decontextualized analysis among social scientists in the postwar period. What is clear is that, in following axiomatics, this was not an unfortunate outcome of increased mathematization but rather a fundamental aspect of it.

The final essay in *Mathematical Thinking in the Social Sciences* was written by Herbert Simon. Another towering Cold War scholar, Simon contributed to numerous areas of research ranging from organization theory to decision-making, artificial intelligence, and information systems.[61] He coined the term "bounded rationality" and won the Nobel Prize in economics in 1978. In 1945, when *Theory of Games and Economic Behavior* was published, Simon printed a glowing review of the book in the *American Journal of Sociology*. "The most important contribution of the entire theory to the social sciences," he noted, was the authors' choice to "construct a formal, mathematical description of a game."[62] For Simon, who believed that the mathematization of sociology was not only desirable but also inevitable, the book exemplified what successful mathematization could be, and he pleaded with his readers to read it and expand upon it.

In 1957, Simon published *Models of Man: Social and Rational*, a collection of sixteen essays he had previously published in journals. The book advanced the case for the use of mathematics in the social sciences on axiomatic grounds. In order to demonstrate the method's utility, in the second section of the book, Simon mathematizes previous theoretical works on social processes in small groups. In the introductory essay to this section, Simon insists that it would be wrong to see these papers as "merely translation," because the mathematical translation itself comprises a contribution to the theory. He then adds, "Mathematics has become the dominant language of the natural sciences not because it is quantitative—a common delusion—but primarily because it permits clear and rigorous reasoning about phenomena too complex to be handled in words."[63] The use of mathematics is especially needed in the social sciences, Simon explains, because social phenomena are highly complex. Expounding further on this point, Simon explains that the mathematization of a body of theory clarifies its concepts by examining "the independence or non-independence of postulates, and in the derivation of new propositions that suggest additional ways of subjecting the theory to empirical testing."[64]

Having constructed a formal model, Simon (like Coleman and Weyl) stresses that the separation between structure and content enables a researcher to reinterpret the model anew. Whereas the model's original construction is inspired by George Homans's *The Human Group*, which considers the relation between "friendliness" and the level of "activity" in groups, Simon explains that the model is also applicable in situations outside of Homans's analysis. For example, it can be used to study the formation of cliques within groups, or perhaps competition between groups. Finally, by renaming the variables, the model can describe the activities of an individual instead of a group, or even government regulation. Simon explains, "Here A would be interpreted as the actual degree of conformity to the regulation, F as the social pressure to conform, E as the effect of formal enforcement activity."[65] Once concepts are no longer defined inductively *but only* within a given system, multiplicity of interpretation follows.

The ability to interpret the model differently, Simon remarks, does not imply that "the psychological mechanism involved in all these situations is identical."[66] Rather, "the underlying similarity appears to be of a rather different character."[67] Simon explains that all of the situations he describes contain a combination of two motivational forces, one external and the other internal. "It is the combined effect," he writes, "of two such motivational forces that produces in each case phenomena of the sort we have observed."[68] Namely, as mathematicians held, axiomatic reasoning helped uncover the "real" underlying mechanisms that governed phenomena. Ignoring the particularities of each example was necessary in order for the researcher to uncover the "true" structural mechanism underlying these diverse phenomena. Contextualized and detail-oriented analysis, so the argument went, could never have accomplished that.

Not everyone was as optimistic about the mathematization of the social sciences. The same year that Simon's *Models of Man* was published, George A. Miller, pioneer in cognitive psychology and original member of the SSRC committee, was ready to offer a more somber analysis of the use of mathematics in his field. In a presentation before the annual meeting of the board of directors of the council in September, Miller noted that too often mathematical models constructed by researchers did not teach them anything new about the underlying phenomenon: "It seems to me that some of the models that have been proposed for the general problem of psychological measurement are more useful in teaching the student how to axiomatize his ideas than they are for any practical problems of measurement in research."[69] Miller was not opposed to the use of mathematics, which, he noted, was continuously increasing. Rather, he wished to emphasize that, in order to be useful, mathematical modeling must offer some insight. Moreover, he worried that in their infatuation with mathematics, researchers would limit themselves only to those areas that could be approached mathematically.[70]

However, even in his more cautious celebration, Miller sought to emphasize that the true power of mathematics was in keeping one's thinking "clear and relevant." In conclusion, he wrote, "I got the impression that many of our colleagues . . . fail completely to see mathematics as *a way of thinking*."[71] For a host of social scientists, mathematics transformed in the years following World War II from an auxiliary tool for research to an indispensable one. This transformation was predicated on foundational changes in the epistemology of mathematics itself, as well as a reevaluation of its explication of physical theories. It also depended on the movement of mathematicians who chose, after the war, to travel in more expanded circles.

Social Axiomatics in Action

Axiomatic thinking, however, was not propagated by mathematicians alone. Probably one of the greatest impetuses for its proliferation in the social sciences came from the publication in 1951 of Kenneth Arrow's *Social Choice and Individual Values*. Indeed, rational choice theory offers one of the clearest examples of axiomatic theorizing in the social sciences. Following the publication of Arrow's book, it became common for certain social scientists to open their books by clearly stating a set of postulates and then proceeding to deduce their theory from them. Arrow developed his axiomatic approach to rational choice theory while working as a researcher at the Cowles Commission for Research in Economics and at RAND, and in the introduction he thanks mathematician J. W. T. Youngs and philosopher Abraham Kaplan for helping him formulate the problem. According to S. M. Amadae, the greatest achievement of Arrow's work was not its formal derivation but how it served to anchor American economic and democratic liberalism against Marxist attacks.[72] It also offered a model for axiomatics.

Arrow's book famously begins by defining rationality as a set of axioms based in set theory. The basis for the work is an assumption that an individual given two choices can always decide between the two (or remain indifferent), and that these choices are transitive (i.e., if you like apples more than bananas and bananas more than oranges, you definitionally like apples more than oranges). The key to Arrow's approach was that by requiring an individual's preference be arranged according to a sequential order, he had no need to introduce any quantitative measurement. Thus, while the book is mathematical in nature, it is not quantitative. There are definitions, proofs, and theorems, but it is symbolic logic and set theory that define the derivations. From this relatively simple set of axioms, Arrow is able to deduce his "impossibility theorem," which asserts that there is no way to aggregate individual preferences into a collective social decision that adheres to every individual's choice. Arrow presents his theory as applicable equally to an analysis of democratic voting, economic decision, or social welfare. This was a direct outcome of his axiomatic approach, as Arrow notes: "One of the greatest advantages of abstract postulational method is the fact that the same system may be given several different interpretations permitting a considerable saving of time."[73] Once again, analytic separation facilitated economy of thought.

Commenting in 1956 on the use of mathematical models in the social sciences, Arrow explained, "It is simply not true that mathematics is useful only in quantitative analysis." Arrow insisted that when his colleagues argued that mathematics was not useful in the social sciences, regardless of its utility in the physical sciences, they were demonstrating their ignorance of modern mathematics: "To the mathematicians or the individuals trained in the spirit of *modern* mathematics, the views just presented [the incompatibility of mathematics in the social sciences] seem to be based on nothing more profound than a misunderstanding."[74] Arrow was one of many social scientists in the postwar period who argued that mathematical symbolism excelled exactly where the messiness of ordinary language failed.[75]

In 1957, this lesson was adopted wholeheartedly by Arrow's student Anthony Downs when he published *An Economic Theory of Democracy*. Downs received his PhD in economics from Stanford University in 1956, and his work is considered one of the pillars of early rational choice theory. In a somewhat revealing statement at the opening of his book, Downs acknowledges that almost everything he formally proves in the work can already be found in Walter Lippmann's book *Public Opinion*, published in 1922. What Downs's work accomplished, as such, was the formalization of these ideas in the language of axiomatics. Downs sets up two axioms for his study: (1) that all citizens are rational in the sense that they seek to maximize utility from governmental action, and (2) that parties are rational in the sense that they seek to maximize votes. From these, he establishes a set of "testable propositions" corresponding to each hypothesis. For example, two derived propositions that he suggests can be tested are "democratic governments tend to redistribute income from the rich to the poor" and "democratic governments tend to favor producers more than consumers in their actions."[76] The derivation itself, however, uses little to no mathematics. Statements are put into symbolic form and propositions are proved, but for the most part this entails logical inferences in *natural language* and some basic graphical representations. Mathematical thinking, or axiomatic thinking, characterizes the work, not mathematical technique.

What marks Downs's work as truly axiomatic is his insistence on a separation between analytic formalism and empirical reality. Downs takes care to account for this feature of his approach. His analysis is built on the assumption of rationality. To be a rational agent, according to Downs, is to be able, in every situation, to clearly order one's preferences and ensure that they are transitive and consistent. However, as Downs is quick to explain, the "rational citizen" is an "abstraction"—in most situations humans are guided by more than one set of considerations. You might prefer the Green Party candidate to a Democratic one, but your friends might pressure you to vote differently. The "model world," Downs acknowledges, "is inhabited by such artificial men," and is hence limited; but it is fair to assume that, on the whole, citizens strive for political welfare.[77] Downs thus instructs the reader to keep in mind the separation between the world his model describes and the world as it is: "The statements in our analysis are true of the model world, not the real word, unless they obviously refer to the latter. Thus, when we make unqualified remarks about how men think, or what the government does, or what strategies are open to opposition parties, we are not referring to *real* men, governments, or parties, but to their model counterparts in the rational world of our study."[78] To verify the model, Downs explains, one must test its conclusions, *not* its assumptions. Only once the formal analysis is complete can judgment begin.

Another foundational book in rational choice, James M. Buchanan and Gordon Tullock's *The Calculus of Consent: Logical Foundations of Constitutional Democracy*, offers another example of how axiomatics spoke more to the separation of formal analysis than to any mathematical content per se. Published in 1962, *The Calculus of Consent* sought to extend economic reasoning into political theory. Buchanan and Tullock postulate "methodological individualism," according to which there is no collective entity one can discuss outside the sum total of the views, needs, and desires of individuals, and in which it's assumed that the average individual "will choose 'more' rather than 'less.'"[79] The writers' interest is in two types of collective decisions: "constitutional," which are determined when a state is first organized, and "operational," detailing the allocation of resources.[80] The analysis includes graphical and symbolic presentations, but despite the "calculus" in its title, the derivations do not follow a formal mathematical proof. Rather, they are presented in plain language. Buchanan and Tullock do utilize ideas from game theory and provide a cost-benefit analysis, but the work does not require high mathematical sophistication from its reader.[81] When, five years later, Tullock published *Toward a Mathematics of Politics*, he was taken to task by none other than Arrow. In a review of the book, Arrow exclaimed, "The title is misleading; there are no developments of any mathematical complexity. The mode of analysis is that of literary economic theory, the logical development being conveyed in words rather than symbols."[82] By 1969, "mathematics" meant a style of reasoning.

Here, again, it is analytic independence that makes the work axiomatic. In the concluding chapter of the book, the two authors explain that "relevant theory is made up of two parts."[83] The first part is the logical model, which, on the basis of postulates and assumptions, derives new consequences. "This sort of theorizing is purely logical in nature and has no empirical relevance in the direct sense."[84] Here, one should recall Veblen's insistence that geometry is similarly divided into two parts: an empirical side, studied by the physicist, and a formal mathematical side. By midcentury, this dual nature of theory had become one of the markers of axiomatic thinking beyond mathematics. Like Downs, Buchanan and Tullock explain that to test the model one must test its predictions. This latter part they call the "operational model."

Many reviewers remarked upon such a bifurcated view of theory. Commenting on Downs's book in the *Journal of Political Economy*, political scientist Martin Diamond explained, "The model touches the real world at two times, just before it is set to working and just after." The model maker started, he continued, "by making certain assumptions about his subject matter (arbitrarily, common-sensibly, in mimicry of other disciplines?—by whatever criterion, apparently the process [was] prescientific)."[85] For Diamond, theorists' ability to isolate their analysis once their initial assumptions were stated was the strength of the approach. Not everyone agreed. In the *American Political Science Review*, Hayward Rogers took Downs to task for his use of axiomatics. Even if we assume that the theorems can be logically derived from the axioms, and that they can be empirically verified, Rogers questioned, "can one then claim, as Downs does," that the axioms can be confirmed? If "neither the axioms nor the theorems are 'interpreted,'" Rogers continued, how can they then be empirically tested?[86] Rogers's uneasiness derived directly from the separation of analytic coherence and empirical verification that axiomatic theory entailed. What explanatory power, he asked, did the model world hold for the "real" word of politics?

Rogers adhered to the older view of theoretical construction, stating that an empirical study of the "rational man" could start with some "guesses at what generalities hold," but that only after those guesses had "been empirically tested and confirmed" could one "deduce further generalizations and test them in term."[87] Induction, not deduction. In a short note, Downs replied to Rogers's critique, defending his approach. He again explained that, while the axioms were unverifiable, the derivation in the model strove to mimic the real world. The goal, he explained, was to deduce conclusions that were testable in reality. He wrote, "These behavioral deductions . . . if empirically verified, tend to confirm the truth of the basic hypothesis."[88] Downs and Rogers fundamentally disagreed on how the idealized model world and the real world related to one another. More broadly, their dispute points to the way in which the adoption of axiomatics forced researchers to interrogate the meaning of theory in the social sciences.

On Theory

One of the questions that plagued axiomatic theorists from the start involved the nature of the theories they were producing: Were they normative or descriptive (and, as such, predictive)? Or were they heuristic? This is where the difference between modern axiomatics as applied in mathematics, physics, and even engineering, on the one hand, and in theories of human sciences on the other, became hardest to square. The separation of analytic formulation from observation had far more significant implications to social scientists than to those in other fields. The key feature of modern axiomatics was that axioms did not need to be inductive. They could be based on observations and experiments, but that was not necessary. For mathematicians, axioms held no truth value outside of the system. Following Hilbert, truth was an attribute of the system as a whole, based on the test of consistency and independence. For many midcentury mathematicians, axioms were not only arbitrary but interchangeable. Several axioms could account for the same theory. The key was that, once an axiomatic system was derived, new theorems could be proved and new connections between separate areas of study could be achieved. When it came to physical theories, the claim that axioms did not need to be verified by experiments was countered by demands that the theory, once constructed, should not only be able to describe physical phenomena but also have predictive value. (A similar emphasis on immediate utility was present in engineering.) Neither of these options were open to social scientists.

Unlike pure mathematicians, social scientists were interested in saying something about the world itself. They were not satisfied by the construction of systems upon systems that, while remaining consistent, could not be verified. Yet most social scientists acknowledged that their theories couldn't be predictive, as theories were in physics. Buchanan and Tullock explained, for example, that physicists also made simplifying assumptions, but were then able to make predictions that could be verified by real world events. "The physical scientist is not, however, dealing with man," they wrote, "and the study of human beings in association with each other introduces a whole set of complexities that remain outside his realm. Social science can never be 'scientific' in the same sense as the physical sciences."[89] Human behavior was simply too messy to be able to perfectly predict the outcome of a group's interaction or an individual behavior in a particular circumstance.

As axiomatic thinking began finding its way into the human sciences, researchers had to account for this difference. Were their theories descriptive? Were they prescriptive, heuristic, explanatory? Isaac has described this reflexive instinct in postwar social scientists as "tool shock." According to Isaac, epistemological concerns "emerged as practitioners of those disciplines struggled to make sense of the dizzying expansion of their tool kit."[90] Here I want to focus on how the adoption of axiomatic thinking (rather than any particular mathematical technique) forced researchers to grapple with the changing meaning of theory in the social sciences, and the relation between observations and analytic formalism. This was true of the most, as well as the *least*, mathematical theories.

In the introductory essay to *Decision Processes*, Robert L. Davis details one of the issues that triggered disagreement and confusion among researchers during a symposium in Santa Monica: the distinction between a theory that is descriptive or predictive and one that is normative. Specifically, the convened psychologists and sociologists found fault in the assumption of "rational man," since the theory "had no apparent relation to observed behavior."[91] This view, however, did not do justice to the work, according to Davis, since "modern formulators" were modest. They did not try to tell people how they *should* act if they were "rational," but rather claimed that if they acted in accordance with the abstract system, they could be considered rational by the theory's definition of the term. Further confusion arose, Davis writes, from the fact that a theory could be both normative and have empirical applications (the original formulation of game theory is given as an example). Hence, in criticizing a theory, one must be clear about which aspect of it one found faulty (the empirical or the normative), "and, of course, both types of argument should be distinguished from any attack on the abstract theory."[92] It was the formulation of their theories on axiomatic grounds that forced researchers such as Davis to create sharp distinctions between the normative, empirical, and formal aspects of their theories. And despite Davis's assertion that "these distinctions are so obvious," it was not always easy to discern where the lines should be drawn.

In *Games and Decisions*, the question comes to the fore when Raiffa and Luce discuss utility theory. As soon as they complete a formal presentation of their axiomatic theory, they are quick to point out that the assumptions in their theory do not necessarily accord with observations. First, the theory is built on the assumption that an individual is faced with an infinite number of pairwise choices, but in any experimental situation only a finite number will be given. Second, the reported preferences of an individual often do not satisfy the axioms' assertion that they are transitive. Third, the theory assumes that an individual can compute the objective probabilities of each action in a given situation. In a normative theory, the authors explain, this last assumption is not questionable, but, they write, "if we are trying to *describe* behavior, it may be unreasonable to suppose that people deal with objective probabilities as if they satisfy the axioms of the calculus of probabilities or that they only cope with situations in which objective probabilities are defined."[93] What, then, can be said about the nature of their theory? Following the physical sciences, the authors explain, it is possible to base a theory on unmeasurable quantities and still be able to derive meaningful conclusions from it.

Raiffa and Luce, however, seem uncomfortable with such a state of affairs, and assert that it would be much better if the "postulates of the model [could] be confirmed." Perhaps, they wonder, there might be some limited situations in which the postulates, or at least a modified version of them, could be experimentally shown to hold. "It may still be an act of faith to postulate the general existence of these new constructs," they write, "but somehow one feels less cavalier if he knows that there are two or three cases where the postulates have actually been verified."[94] This last sentiment makes clear the authors' unease with applying axiomatic thinking in the human sciences. Raiffa and Luce try to find a compromise. Sure, they reason, the postulates are not inductive, but once the model is completed and the analytic framework is established, researchers should strive experimentally to corroborate the underlying assumptions, even if only in a limited number of situations. Such attempts to square the theory with phenomena can be viewed as efforts to find a middle ground between pure axiomatics and inductive social theory. In these efforts, Raiffa and Luce were not alone.

Downs, for example, insists in *An Economic Theory of Democracy* that he is not interested in producing a normative theory. His theory, he explains, is deductive, in that it is built from a set of stated assumptions. However, he explains, "it is also positive, because we try to describe what *will* happen under certain conditions, not what *should* happen."[95] Downs does not claim that his theory is descriptive. His focus is on the notion that it is positive, as it yields propositions that can be tested.[96] It is not predictive per se, but it is verifiable.

Downs makes a further claim. According to him, axioms can be verified if certain of the propositions deduced can be shown to be true. Downs concludes his work by providing two sets of such propositions, and indicates which axioms he has used to derive each. For example, one proposition is "both parties in a two-party system agree on any issues that a majority of citizens strongly favor."[97] Thus, if the proposition can be shown to be true, he explains, the axiom can be said to be verified. Indeed, in the last chapter, this hypothesis is stated to be the "main thesis" of the entire work. Axiomatic reasoning had been stretched thin in its application to political science. For mathematicians, the axioms were completely arbitrary; the test of an axiomatic system was its ability to prove or predict new knowledge. For Downs, the order was reversed.

Buchanan and Tullock, interestingly, often use the word "theory" in quotation marks to note the difference between scientific theory and the sort of theory they have in mind. They also emphasize the explanatory value of such formal constructs. Even if assumptions of economic rationality "are not predominant enough in human behavior to allow predictions to be made, the formal theory remains of some value in explaining *one* aspect of that behavior and in allowing the theorist to develop hypotheses that may be subjected to conceptual, if not actual, testing."[98] This positivist conception of theory was promoted most strongly by William H. Riker. In 1962, Riker published *The Theory of Political Coalitions*. He believed that the greatest failure of "traditional political science" was its overemphasis on gathering information without an adequate attempt to construct a theory of politics. Riker followed a decade later with *An Introduction to Positive Political Theory*, in which he explained the move from inductive to deductive theories as a historical process.[99] Only after empirical details were sufficiently gathered was it necessary to turn to deduction to generate new theories. Riker's historical understanding of the process is most akin to Hilbert, for whom axiomatics became necessary once a theory was already well established. The role of axiomatics was to place existing knowledge on sound foundations. Riker believed that political science had finally entered this mature state. He explained that the adjective "positive" indicated an emphasis on "theory of axiomatics, deductive type."[100] For him, axiomatics and positivist theory were synonymous.

Other researchers, noting that axiomatic theory in the social sciences could not be held to the same standards as in the physical sciences, insisted that it could nonetheless have explanatory power. Anatol Rapoport was another mathematically trained individual whose interests shifted to the social sciences in the postwar period. While Rapoport was interested in game theory, unlike Raiffa and Luce he is mostly known for his work in general systems theory. Rapoport was a student of Rashevsky and earned his PhD in mathematics from the University of Chicago in 1941. His expertise was in mathematical biology, but like his adviser he became interested in applying mathematical models to the human sciences more broadly. In 1958, he published "Various Meanings of 'Theory'" in the *American Political Science Review*, in which he reflected upon the changing meaning of theory in the social sciences in light of increased mathematization. In the article, Rapoport locates much of the difficulty in constructing sociological theory in the act of concept formation. Definitions, he insists, are arbitrary. They are nothing more than conventions on the path toward theory. For too long, Rapoport explains, researchers have assumed that "entities called politics, society, power, welfare, tyranny, democracy, milieu, progress, etc., actually exist, just as cats, icebergs, coffee pots, and grains of wheat exist."[101] The distinction between power and coffee pots, Rapoport continues, is not one of existence ("A 'cat' is no less an abstraction than 'progress,' when you think of it").[102] Rather, it is a question of consensus—what we all agree on when we call something a "cat" or "welfare." It is much harder to come to an agreement on what terms like "democracy" or "power" mean, and it would be futile to try to study these terms if there were no agreement about them.

Like the authors of *The Anatomy of Mathematics*, Rapoport believed mathematics was the way in which such difficulties could be handled. A year later, during a symposium on "sociological theory," Rapoport explained, "Mathematics is the only language we have which is uncontaminated by bias derived from content. Being contentless and independent of specific experience, mathematics is the only 'cosmopolitan' language possessed by man."[103] According to him, this was what made mathematics so useful in the social sciences, but he, too, acknowledged that the meaning of theory as such needed to be reevaluated. Having distinguished between predictive theory in the physical sciences, normative theory in politics, and intuitive or linguistic theory (continental theory), Rapoport offered game theory as an example of a theory that was not predictive and was idealized in its assumption and description of the world ("preferences are never so clear as to be measured in utilities; men are seldom rational"), but was nonetheless valuable.[104] First, according to Rapoport, game theory represented a step forward because it "burst through the framework of thought imposed by physical science."[105] Second, game theory gave rise to new social scientific concepts: "The floodgates have been opened and a torrent of entirely new concepts has rushed by."[106] Third, "game theory . . . [was] of relevance to political science because its fundamental concepts [were] idealizations of what political science [were] about." And fourth, perhaps most importantly, it "distill[ed] the logical *essence*" of political acts.[107] The mathematization of the social sciences simply required a new conceptualization of theory. Strict empiricists might object to such pure theory, he noted, but that was because they did not understand its strength. A physicist, Rapoport concluded, could spend years observing ocean waves, but he would not truly understand the "essential" nature of a wave in motion. This essence could only be reached through logical analysis.

One of the clearest attempts to clarify the various meanings of theory in the social sciences, and one that illustrates how malleable the axiomatic method became as it traveled, came from James Coleman. After he graduated from Columbia University with a PhD in sociology, Coleman spent a short stint at the University of Chicago and then began teaching at Johns

Hopkins University. He became an expert on education, and in 1961 he published *The Adolescent Society: The Social Life of the Teenager and Its Impact on Education*, followed in 1965 by *Adolescents and the Schools*. The first was a study of ten Chicago high schools, with an emphasis on the lives of teenagers and their social circles. Upon publishing this work, he was asked by the National Center for Education Statistics to conduct a study of educational equality in the United States. Published in 1966, *Equality of Educational Opportunity*, or the "Coleman Report," identified the socioeconomic status of a child's parents as the most determinant variable in the success of a child. Throughout the period, Coleman maintained his interest in methodology, and in 1965 he published *Introduction to Mathematical Sociology*, which he dedicated to Lazarsfeld.

In this book, Coleman identifies several uses of mathematics in social scientific theory, from the relatively simple quantification that allows one, for example, to distinguish between individuals belonging to the middle class and the lower class for the purpose of analysis, to the use of latent structure analysis in the measurement of hypothetical constructs such as "attitude" or "norm." Coleman insists, however, that the ultimate aim of mathematics is in theory construction. Quantification by itself, he explains, has no theoretical relevance. "The power of mathematics in empirical science," he writes, "rests in its power as a language for expressing the relations between abstract concepts in a theory."[108] Coleman distinguishes between a "synthetic" theory and an "explanatory" theory. What differentiates the two, he says, is the status of the postulates. Explanatory theories would begin with some known empirical generalization, and then seek to find a set of postulates that would account for such an observed phenomenon. Synthetic theories, on the other hand, would begin by taking a set of empirical generalizations and setting them as postulates, and then deduce further elaborations from them. Testing explanatory theories, Coleman explains, involves checking the derived statements against the initial ones for consistency, and making new predictions (this would be Downs's model). With synthetic theories, confirmation comes from testing the postulates themselves. The two types of theories, Coleman emphasizes, have the same "logical character" and differ only in the information open to observation. A researcher might only be able to observe an individual behavior, and hence can only surmise a subject's internal state. On the other hand, synthetic theories "organize out immediately observable experience into larger systems of action which are beyond immediate observation."[109] Coleman's analysis is an attempt to come to terms with the changing meaning of theory in the increasingly mathematized social sciences in the postwar period.

Among those who sought to move beyond strictly verbal theories and offer more than a collection of observations, a certain consensus emerged that a "theory" took the form of postulates and theorems. However, it was not clear how to relate these theoretical constructions to the real world. Coleman wished to clarify just that. He acknowledged, however, that the distinction between synthetic and explanatory theories was not always clear and could change over time. Moreover, some theories did not follow either archetype: "There are other theories or models which have a very elusive relation to the real world . . . this is not to say, however, that such models are of no use."[110] These latter ones could be simply suggestive, or used as training tools for mathematically inclined researchers. Modern mathematics was predicated on a separation between analytic coherence and observation, which gave rise to new mathematical theories and areas of research. As numbers gave way to relations and structures, it was not only mathematics that transformed, but also mathematization.

It would be foolish to deny the influence of logical positivism on the human sciences in the postwar period. The works of Carl Hempel and Ernest Nagel were influential and used by those seeking to produce formal theories in the social sciences. However, it would also be a mistake to disregard the role of mathematics in determining the scope of theoretical knowledge in the field. For many researchers, mathematics played an integral part in theory construction. Philosophical considerations were important, but when it came to developing a new theory it was mathematics, *not* philosophy, that played a crucial role. This is clear as long as mathematics is not equated with quantification.

From the Sciences to Art

Eleven years after Lévi-Strauss pleaded with his colleagues to recognize that mathematics did not equal numbers, the claim was brought up again, this time not by an anthropologist or a social scientist but by designer and architect Buckminster Fuller. In his contribution to György Kepes's *Structure in Art and Science*, Fuller quoted directly from MIT's Department of Mathematics' annual statement of self-definition: "Mathematics, which most people think of as a science of number is, in fact, *the science of structure and pattern in general*." Fuller explained, "[this definition] agrees comfortably with my definition of the word 'structure'—structure is not a 'thing'—it is not 'solid'"[111] Fuller was not the only one who turned to modern mathematics for a theory of design and arhitecture in the postwar period. As Theodora Vardouli has demonstrated, during the 1950s and 1960s a diverse group of practitioners, including Christopher Alexander in the United States, Sir Leslie Martin in England, and Yona Friedman in France, adopted the structuralist conception of mathematics in order to transform architectural *theory*. In particular, they appealed to mathematics not only in reduced form as geometry or measurement (although this too was the case), but also as a *mode of thinking*. As Vardouli argues, architects turned to modern mathematics as a tool with which to "produce rigorous theories of architectural design—ones that would be built on rationally verified principles instead of empirical traditions, would be applied across multiple design situations, and would help architects relate the physical form of building with its social or environmental performance."[112] Mathematics was a tool with which to probe the underlying *theory* of design.[113] The modern conception of mathematization was not limited to the human sciences.[114]

Friedman's work is the clearest example of axiomatics' infinite flexibility. In *Toward a Scientific Architecture*, Friedman dedicates an entire section of a chapter on method to a theoretical discussion entitled "Choosing an Appropriate Axiomatics."[115] The method Friedman describes is not identical to that of mathematicians, but the similarities are nevertheless remarkable. The axioms, Friedman insists, must meet three conditions, which Hilbert had enumerated seventy years earlier: they must not contradict one another, they must not be redundant, and they must be complete. Friedman then translates this axiomatic approach into a graphical representation before concluding that, for all intents and purposes, axiomatic systems in architecture should consist of—at most—three axioms. Having described the "abstract reasoning" behind the methodology of axiomatics, Friedman applies it to the work of the architect. The first axiom states that "the work of architects and planners produces enclosures—separations in space"; the second asserts that "a separation of space (an enclosure) cannot be the work of architects and planners if it does not have at least one access"; and the last declares that "in a system of spatial separations there must be at least one enclosure that differs from the others in some respect, whether as a result of physical qualities or of others."[116] Friedman acknowledges that these axioms may appear "trite," but the point, he insists, is that "they allow for any solution imaginable."[117]

That the modern conception of mathematics was adopted by architects and designers on the one hand and postwar human scientists on the other would not have been of great surprise to mathematicians themselves. As I discuss in chapter 4, throughout the postwar period pure mathematicians turned to axiomatics to claim that their field belonged equally among the arts and the sciences.

CHAPTER FOUR

Creative Abstraction

Abstract Art, Pure Mathematics, and Cold War Ideology

Abstraction is an old story with the philosophers, but it has been like a new toy in the hands of the artists of our day.
—Robert Frost, "The Figure a Poem Makes," 1939

In 1963, an applied mathematician named Bernard Friedman found himself in a room full of science writers, trying to explain the difference between pure and applied mathematics. He noted that most mathematicians employed by large corporations did not consider their work to be mathematics. John Osmundsen of the *New York Times* asked Friedman to explain what he meant. "Most Madison Avenue artists do work they don't consider art," Friedman responded by way of analogy. He continued, "It's the difference between Jackson Pollock and Norman Rockwell . . . The students today find the same excitement in pure mathematics that artists do in Abstract Expressionism, because, you see, neither study has any necessary connection with the real world."[1] Surmising that the journalists in attendance were more likely to be familiar with the work of Pollock than with the output of modern mathematicians, Friedman used abstract art to illustrate that the latter also sought inspiration for their work outside the physical world.[2] Only four years earlier, *Life* magazine had published an essay, rich with photographs, declaring that Abstract Expressionist artists were "the most influential painters in the world today."[3] Art offered Friedman a convenient analogy with which to explain the peculiarities of high modernist pure mathematics to a group of outsiders.

Many postwar mathematicians drew a similar connection between art and mathematics. Four years after Friedman, in another attempt to describe the nature of his field to lay people, mathematician Paul Halmos also reached for an analogy. Halmos, who had immigrated to the United States from Hungary at thirteen, was well-known for his illustrative articles and monographs.[4] An accomplished mathematician and a lucid writer, he would later win several prizes for his expository work and serve as chair of an American Mathematical Society (AMS) committee dedicated to setting style guidelines for mathematical writing. In 1967, in front of a crowd at the University of Illinois, Halmos struggled, like Friedman before him, to explain what modern mathematics was all about. Halmos noted that giving a broad audience a feeling for what mathematics *is* and what mathematicians *do* without in the process teaching them some mathematics was practically impossible, but he took on the challenge. Instead of providing examples of mathematical theories, he discussed the origin of mathematics, the nature of the mathematical community, the idea that mathematics is a language, and the distinction between pure and applied mathematics. The thrust of his speech, however, was dedicated to the assertion that, more than simply being analogous, mathematics *was* an art. His goal was to convince the audience that pure mathematical research should be thought of as a creative pursuit.

To do so, Halmos settled on a familiar set of comparisons. Music, he explained, offered one point of connection: neither musicians nor (pure) mathematicians feel any need to justify their work on pragmatic grounds.[5] Literature offered a second analogy; like math, it was irreducible to its elementary parts: "Literature is more than reading and writing, and mathematics is more than figuring."[6] Moreover, Halmos added, while the source of literature was human life, literature was "not the life it comes from." The best analog, he concluded, was painting: "The origin of painting is physical reality, and so is the origin of mathematics—but the painter is not a camera and the mathematician is not an engineer."[7] He continued, "Asking a painter to 'tell a concrete story' is like asking a mathematician to 'solve a real problem.' "[8] Halmos expanded on Friedman. Not only were modern artists and modern mathematicians connected in their opposition to physical reality, but neither community could justify or account for its work via a clear utilitarian metric. Mathematicians were not driven by a desire to solve real-world problems, just as artists were not driven by pragmatism. Halmos, who a few years later would publish an article titled "Applied Mathematics Is Bad Mathematics," believed pure mathematics to be superior to applied mathematics for this reason. The power of mathematics, in his view, was its separation from reality.

Halmos did not turn to art (and, more specifically, to painting) as an analogy. His claim was stronger than that: he believed that mathematics was itself an art form. He concluded, "[Mathematics] is a creative art because mathematicians create beautiful new concepts; it is a creative art because mathematicians live, act, and think like artists; and it is a creative art because mathematicians regard it so."[9] Friedman and Halmos were not outliers. In speeches, articles, letters, and even national reports, mathematicians time and again drew comparisons between their pursuits and those of artists. Some mathematicians argued that their work and art were analogous activities. Others, like Halmos, insisted that mathematics was an art. Indeed, in the postwar period, art, not science, became the most common comparison for pure mathematics. What, then, might one make of such claims? What sort of rhetorical, political, or symbolic work did mathematicians achieve—or hope to—by aligning their research with art? What do such homologies reveal about the nature of mathematical research at the time? And do they point to a substantive connection between mathematics and the arts at midcentury?

I argue that such proclamations can be understood on three levels, which this chapter will address in turn. First, and most simply, mathematicians' self-conscious association with art during this period was undoubtedly an attempt to reposition the field in the wake of great political and social transformations. Almost as soon as World War II ended, mathematicians began to worry that utilitarian demands would dominate the development of the field, and warned that political considerations might curtail research. These two concerns drew mathematicians to art. To them, art, and *modernist* art in particular, symbolized everything that science had lost in recent decades, namely pure and disinterested research. Art offered the only real corrective to the postwar utilitarian spirit. Claiming pure mathematics as art was thus an attempt to similarly position their work outside of the demands of the day.

That mathematicians believed abstract art to be isolated from the political realities of the time does not mean, of course, that this was necessarily the case. Historians of art have long debated the degree to which midcentury artistic abstractions were (or were not) a reaction to the Cold War.[10] Serge Guilbaut has famously argued that, in insisting its art was apolitical, both the US government as well as the art world—artists, critics, and collectors—used the work of Abstract Expressionists to advance Cold War agendas.[11] The US government presented the individualist and nonideological bent of Abstract Expressionism as a testimony to the superiority of American freedom over the Soviet totalitarian regime.[12] The success of this rhetoric explains why midcentury mathematicians saw in the period's art a symbol of free and nonideological inquiry, regardless of the claim's ultimate veracity.

Second, mathematicians appealed to art as an epistemological double, as is evident in their uniquely modernist conception of art. When they claimed mathematics as a creative pursuit, it was not, as Friedman noted, the paintings of Norman Rockwell that mathematicians had in mind, but rather the work of contemporary abstract artists. Mathematicians turned to abstract art because they felt that nowhere else was the turn to abstraction so transformational as it was in these two disparate fields. As Robert Motherwell exclaimed about abstract art during a 1951 Museum of Modern Art symposium, "How many rejections on the part of her artists! Whole worlds—the world of objects, the world of power and propaganda, the world of anecdote, the world of fetishes and ancestor worship."[13] In no other field of study was the break between the world and the work so complete. To be sure, modern physical theories such as relativity and quantum mechanics were more abstract than classical physics, and the modern social sciences were more theoretical than they had been in the past, but these fields still fundamentally revolved around the lived, material world. Human behavior as a generalized concept might be a fiction authored by social scientists, but motivating that fiction was a better understanding of flesh-and-blood people. Similarly, physicists, even if they believed energy to be nothing but an abstract concept, sought to discover natural laws about the behavior of space and matter. Only artists, mathematicians believed, had abdicated the material world as completely and as forcefully as they had. As Friedman put it, "Neither study has any necessary connection with the real world." This is what lies beneath postwar mathematicians' compulsive promotion of themselves as akin to artists.

I thus read mathematicians' proclamations alongside those of the artists and critics of the time who, like the mathematicians, strove to make sense of their abstractions.[14] Not only were the two groups similarly concerned with the relation between their work and the world, they also developed similar strategies for justifying their output to themselves and others.[15] By midcentury, both artists and mathematicians were committed to aesthetic autonomy, and they realigned abstraction as both a methodological imperative describing the work itself and with withdrawal (i.e., abstraction) as a political stance characterizing the context of its production. This set the postwar bond between art and mathematics apart from a longer history of cross-pollinations between art and mathematics.[16] Numerous scholars have noted how modern mathematical ideas influenced—and in turn were impacted by—developments in the visual arts, architecture, and literature.[17] In this chapter, however, I am less interested in tracing any direct influence. Postwar mathematicians did not emphasize the way mathematical ideas and concepts impacted the arts. For them, what was important was that mathematics was akin to art as a creative practice, and that this artistic conception was rooted in the break between mathematics and reality.[18] Mathematics' likeness to art was not based in some long-held aesthetic concept such as symmetry, pattern, or ratio, but existed because both were imbued with an unparalleled freedom of creation.

Finally, such discursive analysis of mathematicians' commitment to an artistic image of their field helps account for the ubiquity and appeal of abstraction at midcentury, across a wide variety of intellectual activity. Mathematics lays bare the contradictory, and hence productive, nature of abstractionist thought. Midcentury mathematicians situated their aesthetic vision of mathematics in axiomatics. That is, mathematicians argued that *the same methodology* that accounted for the growing applicability of mathematics was also responsible for its artistic side. The utilitarian and the idealistic were not contradictory, but rather two sides of the same coin. In this way, abstractionist thought, understood as relational and semiotic, proliferated during the Cold War. It enabled a diverse set of actors, from artists and sociologists to literary critics, to claim their work as nonideological by elevating method above substance. Whether it was a theory of art or a new theory of human behavior, abstractionist thought putatively isolated intellectual activity from politics.

Abstractionist thought had its roots in the broader transformations underway at the turn of the century. Postwar, however, abstraction's inherently contradictory nature fueled its persistence and proliferation throughout the Cold War—and nowhere is this dynamic more visible than in mathematicians and their talk.

A Humanistic Pursuit

As early as 1943, Marston Morse began worrying about World War II's potentially disastrous effects on mathematics. "When the war is over and guns and bomb sights can be forgotten, what will happen to mathematics?" Morse opined in the pages of the *Scientific Monthly*. "Unless mathematics is somehow associated with the humane studies and with philosophy," he insisted, "its greatest values will be obscured and forgotten."[19] The only way to protect the field, Morse believed, was to instill in students a recognition that mathematics was "an art as well as a science."[20] Morse did not object to the pursuit of mathematical research for military purposes. On the contrary, he both participated in defense research and, as one of the most distinguished American mathematicians of the time, advocated for the integration of mathematicians in war-related scientific work (chapter 2). Indeed, Morse's words above appeared in the conclusion of an article titled "Mathematics and the Maximum Scientific Effort in Total War," in which he called upon the scientific elite to take full advantage of the nation's mathematical talent. Still, he wanted to ensure that as far as *pure mathematics* was concerned, wartime applications were an aberration and not the new norm. Mathematicians should turn their attentions to new fields, he suggested, such as operations research, computing, and information theory, as well as old fields such as ballistics and aerodynamics. But they should be careful not to abandon research in pure mathematics.[21]

Griffith Evans was similarly concerned. In a letter to Marshall Stone in 1942, Evans wrote, "As far as the future position of mathematics is concerned, I hope that the war will not make any great difference. Although the practical value of mathematics is of great importance, its value as one of the humanities is equally so. I hope that both aspects will be kept before the public."[22] Pure mathematicians were eager to participate in war research and were enraged when they believed their efforts had been rebuked by the nation's scientific elite. But they simultaneously feared the pragmatic implications of these new alliances. Even Richard Courant, who advocated forcefully on behalf of applied mathematics, worried during the war that the attention paid to mathematics by military agencies might skew the development of the field. In a 1944 letter to Ronald Kline, he noted, "There seems to exist a danger now that an artificial distinction between applied and pure mathematics will be forced upon us from outside and this distinction might manifest itself in the form of one-sided material and moral support."[23] The only way to make sure this did not happen, Courant added, was to stress the "position of mathematics in between the natural sciences and the humanities."[24] A belief that the "proper" growth of the field required mathematics to remain at least in part a humanistic endeavor was almost universal among mathematicians.

Morse was concerned with the future not only of mathematics but also of liberal education. Mathematicians, he argued in 1943, "must not leave it exclusively to the humanists and philosophers to integrate the knowledge of the day and give it spiritual unity."[25] Too often, he added, mathematics was valued for its power of precision, but such a view was limited: "To give play to the imagination, to *create* and form ideas, to have a mastery of language and logic, to have that *freedom* that comes from recognition of dogma, and the open acceptance or rejection of an axiom *at will*, to recognize how much one does not know, or cannot know; these things are all in mathematics."[26] In claiming the place of mathematics in the humanities, Morse was tapping into a long and robust tradition. Medieval scholars situated the quadrivium of arithmetic, geometry, music, and astronomy as one of the liberal arts, cementing the place of mathematics in humanistic learning. In the early modern period, as Matthew Jones has demonstrated, René Descartes, Blaise Pascal, and Gottfried Leibniz all argued that the proper domain of mathematics was not in its various applications and uses but in cultivating intellect and virtue.[27] In Victorian Britain, as Joan Richards has shown, mathematicians like Augustus De Morgan held that mathematics was the "most quintessentially humanistic subject."[28] Finally, in nineteenth-century Germany, as Lewis Pyenson has argued, pure mathematicians avowing neohumanist values sought to fend off school reforms that sought to emphasize the applicability of mathematical knowledge.[29] It is this history that Morse worried could come to an end.

Morse's plea was made at a time when the militarization of the sciences and mathematics were taken to be at their height. American mathematicians were involved in research during World War I, but the scientific mobilization prompted by the Second World War was unmatched. Mathematicians truly believed that the "purity" of mathematics was endangered, and that their discipline's membership among the humanities was coming to an end. This is what motivated postwar mathematicians to swing so forcefully in the other direction, claiming math as art. The assertion that mathematics cultivated the intellect was prominent in school reform, as Christopher Phillips has shown, but this was not the case that pure mathematicians were making.[30] For them, mathematics was associated with freedom. Morse's image of mathematics was undoubtedly a modernist one. Like his IAS colleague Oswald Veblen, Morse held that axiomatics was a powerful tool against dogmatic beliefs. Axiomatics, and the freedom afforded by it, was what made mathematics an invaluable part of liberal education, not its common association with precision and logic.

During the war, Morse's case for humanistic mathematics was driven by a (mistaken) belief that once the war was over, the "social and human studies" as well as "the arts and philosophy" would reign supreme in academic circles. Ensuring that mathematics remained prominent in the future, Morse believed, depended on the cultivation of its humanistic side.[31] Morse recognized his error in the late 1940s, but he did not stop advocating for his vision of a humanistic mathematics. Rather, he began insisting that it was a crucial antidote to the utilitarian conception of science brought on by scientists' enduring alliance with the federal government and the military. In 1948, Morse was invited to deliver a speech at Kenyon College on "Science and the Library." He took the opportunity to rail against what he took to be the dominant spirit of scientific materialism: "Activity for activity's sake, computation without understanding, statistics as an end, gadgetry instead of science—these things are marks of today."[32] Morse did not condemn science as a whole—only its current incarnation. He wanted to reclaim a purer vision of science encompassing the pursuit of wisdom and understanding rather than the control and manipulation of things.[33]

Other American scientists were similarly concerned that the US scientific community's postwar entanglement with the military threatened a "purer" image of their field. As Jessica Wang has shown, geophysicist Merle Tuve lambasted a perceived postwar conflation of technology and science and sought to reaffirm a vision of pure scientific research.[34] However, as Wang writes, Tuve found it increasingly difficult to square his ideal of science with the realities of postwar funding structures and the demands of Cold War politics: "Tuve's assertions about the virtuous qualities of science and scientists identified technology, military instrumentality, and large organizations as not representing the true nature of science, while simultaneously conceding that they were what science might be turning into, and perhaps even had already become."[35] Morse similarly hailed pure science, but his emphasis was, for the most part, on pure mathematics. The difference is meaningful. Pure mathematicians were happy to retreat to the ivory tower in the aftermath of World War II. While they began to rely more heavily on federal funding for their travel, summer institutions, and publications, they did not, for example, require expensive machinery.[36] For Morse, pure mathematics represented everything that science had lost. It was the last gasp of a vanishing scholarly world.[37]

Throughout the postwar period, many mathematicians offered pure mathematics as the ultimate retort to scientific militarism. The development of the atomic bomb loomed large in their minds, and many worried that military needs were colonizing scientific and mathematical research. They maintained that mathematics had always held (and should always hold) equal footing with the humanities and sciences, and that pure mathematical research represented pure inquiry in its most celebrated form. When Marshall Stone delivered the annual American Mathematical Society Gibbs Lecture on Mathematics and the Future of Science in 1956, he began by noting that, unlike some of his colleagues, he did not object to or even look down upon the application of mathematics. (Undoubtedly, he and his audience both had Bourbaki in mind.) Referencing a recent article by Yale University president Alfred Whitney Griswold, Stone suggested that it was important to distinguish between a liberal education and a utilitarian one: "I hold . . . that utility alone is not a proper measure of value, and would even go so far as to say that it is, when strictly and short-sightedly applied, a dangerously false measure of value."[38] In mathematics especially, he added, "the adoption of a strictly utilitarian standard could lead only to disaster."[39] To further support his position, Stone quoted from an article published two years earlier by Alan T. Waterman, the director of the National Science Foundation at the time, in which Waterman claimed that mathematics "bridges the gap, real or imaginary, which exists between the sciences and the humanities."[40] In the article, Waterman argued that despite its numerous applications, mathematics had never lost its "scholarly aura." Now Stone was calling upon his colleagues to ensure this would remain the case in the future.

Three years later, historian of mathematics Carl Boyer made a similar argument in the pages of *Science*. Mathematicians, he wrote, support new academic demands for mathematical training, but "they look askance at the motives." Echoing Morse, Boyer insisted that mathematical education was as beneficial for the engineer and the scientist as for the philosopher. "Especially, surrounded as we are by pressures of immediacy and expediency, it is necessary to look beyond the caricature of the mathematician as a glorified calculator and to appreciate the part of pure mathematics."[41] Pure mathematics, in this view, offered a corrective to the overwhelming push toward applied research.

In 1949, Stone resigned from the Atomic Energy Commission (AEC) Fellowship Board in response to the AEC's request that all fellowship holders take a loyalty oath and sign a noncommunist affidavit. As Jessica Wang explains, Detlev Bronk, chair of the National Research Council, and Alfred Newton Richards, the president of the National Academy of Sciences, accepted the new fellowship qualifications as a compromise, fearing that Congress would otherwise demand each fellow undergo a full-fledged FBI investigation. Most members of the NAS were troubled by the new policy, but, as Wang writes, they yielded without much protest.[42] Not Stone.[43] The only member to officially resign rather than take the oath, he took to the pages of *Science* to explain his reasoning. "Basically, the issue here is one of political freedom," he wrote. "The denial of educational rights or privileges to a citizen who would be eligible for them were it not for his failure to measure up to some arbitrary political test is a clear violation of the principles upon which our republic was founded."[44] A strong believer in civil liberties, Stone compared the new policy to a case that his father, Supreme Court justice Harlan F. Stone, had heard nine years earlier. In *Minersville School District v. Gobitis*, two Jehovah's Witness children refused to salute the US flag during a daily school exercise and consequently were expelled. When the case was argued before the Supreme Court, Justice Stone wrote the sole dissenting opinion, in which he claimed that forcing a child to salute the flag would violate his liberty and freedom as guaranteed by the Constitution.[45] Three years later, in a similar case, the court reversed its original position and adopted Justice Stone's earlier position. Marshall Stone now averred that his father's opinion "should be read by anyone interested in the present situation."[46]

Stone was not the only mathematician to sound an alarm in 1949. Mathematician Irving Segal, for example, wrote a letter to the secretary of the AMS urging him to bring the issue before membership consideration. Warning that the recent loyalty oaths attached to government fellowships would lead professional science in the US the way of Russia, Siegel noted, "I believe that the society should also insist that government fellowships in pure science should be granted on the basis of professional competence and promise alone, and that political belief or affiliation should be considered immaterial."[47] When the Board of Regents of the University of California instituted a mandatory loyalty oath for all university employees the following year, the Council of the American Mathematical Society, like other professional organizations, passed a resolution condemning the university for "imposing arbitrary and humiliating conditions of employment on the faculty."[48] Loyalty oaths had existed in state legislation since the 1930s, but in the postwar period they increasingly began to include specific language confirming that employees did not subscribe to specific beliefs or organizations.[49] Three months after the council approved its first resolution, it voted to follow up with an even stronger condemnation halting any mathematical meeting on a University of California campus until conditions changed. Before the resolution was passed, some members of the society voiced their disapproval, and a committee was formed to study the case and establish general procedures to follow in the future. Stone was appointed to chair the committee.

Stone, who used to read and comment on his father's court opinions, wrote the committee's final report, which included both a legal analysis of the society's mandate as specified in its bylaws and a legislative history of the impasse in California. Stone concluded that the society not only had a legal right to act but that, in the current climate, it must. "One must recognize," he wrote, "that the social transformations taking place in our times tend to integrate the professional activities of mathematicians ever more closely into a scheme of organization in which decisions affecting the professions are progressively centralized and increasingly entrusted to politicians and administrators."[50] A strong professional organization, Stone insisted, was never so needed. The California loyalty oath was not the first controversy the AMS found itself having to weigh in on. That same year, the council instructed its policy committee to join the American Association of University Professors in investigating the case of mathematician Lee Lorch, who had been denied reappointment by the Pennsylvania State College on account of his political activism.[51] Lorch was a principled opponent of racial discrimination and had previously been dismissed by the City College of New York for trying to desegregate Stuyvesant Town. In consultation with Morse, who chaired the policy committee, Stone forwarded the names of three men "of well-known liberal tendencies" to join the investigation. These cases convinced Stone that mathematicians could and should act to ensure academic freedom.

In his final report, Stone argued that recent history illustrated the increased threat to civil liberties that occurred in times of national stress or emergency: "We believe it would be unrealistic to expect the current wave of loyalty oaths and political tests to subside so long as the tension between the United States and Russia, with its far-reaching political implications, continues."[52] All the experts agreed, Stone added, that tension between the two countries could last for decades to come, and in the meantime, the political tests that policymakers were devising seemed poised to become increasingly legally sophisticated. In the coming years, those holding "unorthodox opinions" would therefore find themselves under increased scrutiny and pressure. Stone wrote, "An important factor, which learned professions will ignore at their peril, is the growing dependence of our educational and research institutions upon public, particularly federal, funds."[53] What Stone and many of his mathematical colleagues feared was a future in which federal and military interest in mathematics came not only with utilitarian expectations but also with conditions as to who could conduct research. Academic freedom, they believed, was in great jeopardy.

Mathematicians were not alone among professional organizations resisting perceived and actual external incursions upon academic freedom.[54] However, in turning to modern art to make a case for mathematical autonomy, their response was distinguished from that of other scientists.

A Creative Pursuit

"Creative work in this field," explained mathematician Adrian Albert in 1960, "has the same kind of a reward for the creator as does the composition of a symphony, the writing of a fine novel, or the making of a great work of art."[55] Such claims were made by numerous mathematicians in the 1950s and 1960s, but no one made the case more forcefully than Morse. Throughout the postwar period, Morse produced several lectures and articles advancing a humanistic ideal of mathematics as art while condemning the utilitarian view of science. He did so not only in front of mathematicians and scientists, but to audiences of artists and their critics as well.

In 1950, Morse was again invited to Kenyon College, this time to participate in a conference in honor of Robert Frost. At the time, Kenyon was a hotbed of New Criticism, which dominated American literary criticism during the midcentury. Critics such as John Crowe Ransom, Cleanth Brooks, and Allen Tate advanced a formalist approach to reading literary work, especially poetry. They believed that consideration of such elements as historical and cultural context, an author's intentions, and readers' responses should be eliminated from literary analysis, which instead should focus solely on the structural elements of a work (e.g., rhyme, meter, and plot). Despite receiving heavy criticism in later years for imitating the sciences (more on this soon), advocates of New Criticism promoted literature as the only response to what they considered an increasing bureaucratic rationalization of American life. This attitude might explain why Morse was invited to Kenyon multiple times. His denunciation of contemporary scientific thought aligned neatly with the advocates of New Criticism, despite their disparate training. Morse himself was notably no stranger to poetry and an avid reader of the *Kenyon Review*. At Princeton,

Morse played Mozart violin sonatas with Allen Tate, and his wife attended classes by Tate's student Richard P. Blackmur.[56]

Befitting his audience, Morse titled his paper "Mathematics and the Arts." Unlike his IAS colleague Herman Weyl or his teacher George Birkhoff, Morse did not focus on the nature of symmetry or an analysis of geometric patterns in his discussion of the bond between mathematics and the arts. He did not claim that mathematical formalism could offer a theory of aesthetics, as Birkhoff had, or repeat Weyl's assertion that mathematical analysis could explain the power of symmetry in art, organic nature, and inorganic matter.[57] Rather, his claim was that mathematicians gain the same feeling and satisfaction from their work as artists do from theirs, and that both pursuits were driven by freedom of thought. The *New York Times* reviewer Donald Adams was so impressed by Morse's comments that he reported, "The most interesting contribution to the conference was a paper by the eminent mathematician Marston Morse."[58] Morse must have known his talk was well received: upon his return to the IAS, he declared to fellow mathematician Raoul Bott that he had "brought the house down."[59]

Morse suggested that the basic affinity between mathematics and the arts was spiritual and psychological. Arguing from history as well as personal experience, Morse maintained that discovery, in mathematics as in art, was a matter of intuition. He wanted to convince his audience that while in popular accounts mathematics was often equated with deductive and logical reasoning, this was a very limited view of the discipline. The essence of mathematical practice was not writing down a complete proof, but rather all the work that preceded it (a large part of which, he maintained, was done unconsciously).[60] To drive his point home, he quoted Frost's "The Figure a Poem Makes": "I tell how there may be a better wildness of logic than of inconsequence. But the logic is backward, in retrospect, after the act. It must be more felt than seen ahead like prophecy."[61] Morse insisted that the same process could describe mathematical discovery. Writing down a logical proof was done only in retrospect, second to the creative act. Like the artist, the mathematician was guided first and foremost by beauty. The mathematician "wishes to understand, simply, if possible—but in any case to understand; and to create, beautifully, if possible—but in any case to create."[62] Morse's emphasis on creation is telling. Crucial to mathematicians' claim to art was the idea that axiomatics refashioned them as *creators*.[63]

Morse was not alone in his thinking. Mathematicians adhered to the idea that, rather than describing the world around them, they brought to life the mathematical realms they studied.[64] In a letter to his philosopher sister Simone Weil, André praised the importance of axiomatics: "When I invented (I say invented, and not discovered) uniform spaces, I did not have the impression of working with resistant material, but rather the impression that a professional sculptor must have when he plays by making a snowman."[65] When they became the building blocks of mathematics, axioms refashioned mathematical research as a dialectical play between freedom and constraint.

In his 1953 autobiography, Norbert Wiener similarly focused on the artistic qualities of mathematical research. Noting, as was common at the time, that the rewards mathematicians gained from their work were similar to the rewards of artists, he added, "To see a difficult uncompromising material take living shape and meaning is to be a Pygmalion, whether the material is stone or hard, stonelike logic. To see meaning and understanding come where there has been no meaning and no understanding is to share the work of a demiurge."[66] Likening the work of mathematicians to a godlike pursuit was an analogy that axiomatics made possible. The arbitrariness of the axioms necessitated that mathematical creations be brought to life by mathematicians. Again, the analogy to art was a modern one—what the abstract artist creates can never be reduced simply to what he observes in the world. Hans Hofmann, who was both a respected artist and a teacher among Abstract Expressionists, defined the artist as "an agent in whose mind nature is transformed into a new *creation*."[67] In his teaching, he explained that creation "is not a reproduction of observed fact."[68] Like mathematicians who were content proclaiming that nature was itself the source of all mathematics at the same time that they insisted that pure mathematics was completely separate from the world, Hofman nonetheless held that nature was the source of all artistic inspiration.

As Linda Dalrymple Henderson has shown, in the 1950s both artists and scientists began pointing to creativity and freedom in discussions of their work.[69] Especially in the aftermath of C. P. Snow's popularization of the Two Cultures, scientific humanists used freedom and creativity to argue that, from an experiential perspective, the sciences and the arts were not all that different.[70] Morse was among the first scholars to insist on the place of creativity and freedom in scientific practice.[71] However, more than demonstrating that art and science were motivated by freedom and imagination, Morse turned to art in his attempts to protect science from its postwar utilitarian incarnation. Toward the end of his Kenyon address, he exclaimed that he was often shocked when he listened to students talk about science: "The science that they speak of is the science of cold newsprint, the crater-marked logical core, the page that dares not be wrong, the monstrosity of machines, grotesque deification of men who have dropped God, the small pieces of temples whose plans have been lost and are not desired, bids for power by the bribe of power secretly held and not understood."[72] This view of science, he insisted, was not one he held: "I shun all monuments that are coldly legible. I prefer the world where the images turn their faces in every direction, like the masque of Picasso. It is the hour before the break of the day when science turns in the womb, and, waiting, I am sorry that there is between us no sign and no language except by mirrors of necessity. I am grateful for the poets who suspect the twilight zone."[73] That Morse refers to the work of Picasso is telling. The art Morse had in mind was not one of mimetic accuracy; rather, what made art powerful was its ability to bring to life, and grapple with, those elements of the world that could not be reduced to simple representations.

When *The Bulletin of the Atomic Scientists* contacted Morse eight years later with a request to republish his talk, he agreed, but insisted on adding a short note. In it, he explained that one of the invited speakers was the poet and novelist Leonard Strong. According to Morse, after Strong finished reading one of his poems, he placed his hand on his head and said, "Science is here," then placed his hand on his heart and said, "and poetry is here."[74] Feeling that Strong's characterization of science was misguided, he immediately set to rewriting the concluding paragraphs of his speech. Poetry and mathematics were not at odds with one another, he argued; reason played an important role in mathematical research, but it was second to intuition and understanding. Of Strong, Morse added, "I have no doubt he had in mind the opening of the atomic era," and this was exactly the image of science against which Morse protested.[75] Throughout the 1950s, Morse delivered several more speeches arguing for the bond between mathematics and art.

By the 1960s, the assertion that mathematical research was a creative art had made its way into an official national report on the state of mathematics. Whereas the report focused on the utilitarian view of mathematics, the final section was devoted to the "nonutilitarian view" of the field. It would be a mistake, the section asserted, to suggest that all "society's demands on mathematical science [were] utilitarian—a vivid untruth."[76] Even if mathematics had no utility, society would continue to value the field for its "intellectual strength and beauty." The report also asserted, "The appeal that mathematics has for its practitioners, in particular research mathematicians, is largely aesthetic. The joy experienced in learning and in creating mathematics is akin to that associated with art and poetry."[77] It might seem baffling that, having spent more than two hundred pages extolling the countless uses of mathematics, the authors of the report chose to conclude by affirming the discipline's aesthetic quality. However, this was fundamental to mathematicians' claim to the humanities.

If this were the extent of such proclamations, then the analogies mathematicians drew between art and mathematics could most likely be dismissed as nothing more than rhetorical. As Jamie Cohen-Cole notes, in Cold War America social critics promoted freedom, autonomy, and creativity as necessary traits in the fight against authoritarianism.[78] Mathematicians, conceivably, might have simply adopted the language of the time to try to protect the field. While it is certainly true that comparisons between mathematics and art were frequently drawn in service of supporting pure research, the arguments extended beyond talk. Mathematicians believed free inquiry was conditioned not only by the elimination of political influence and pragmatic demands but also by the subject matter itself. In other words, true freedom could only be achieved by turning one's eyes away from the world. This is what midcentury mathematicians emphasized in their turn to modernist art. In doing so they adopted a particularly modern definition of abstract art as nonrepresentational: The artistic quality of a mathematical theory was proportional to its abstractness. The midcentury adherence to creativity became inherently bound to axiomatics.

Aesthetic Autonomy

In 1984, mathematician Peter Lax conducted a survey on the development of mathematics in the preceding decades. It was only in the past century, Lax remarked, that "the bold proposal to cut the lifeline between mathematics and the physical world was put forth."[79] Lax decried the move, holding that for mathematics to continue to flourish it must maintain some footing in physical reality. His objection did not end there. Breaking the bond between mathematics and reality, he explained, was not only "wrong-headed," but it also "raise[d] profound philosophical problems about value judgment in mathematics. The question 'What is good mathematics?' [became] a matter of a priori aesthetic judgment and mathematics [became] an art form."[80] Since pure mathematical research at midcentury was not motivated by a desire to discover natural laws, the only way to distinguish between two theories, according to Lax, boiled down to "aesthetic judgement." Lax thought that aesthetic considerations did have a role to play in mathematical research, but not a dominant one. For him, mathematics could be *like* art, but it should not *be* art, or at least not abstract art.[81]

As should already be clear, for many pure mathematicians, claiming that mathematics was an art was not a critique but a point of pride. It was a way to access, or give voice to, an element of mathematical practice that was often left unacknowledged. Pure and applied mathematicians might have disagreed about how far the artistic quality of mathematics should be pursued, but they agreed that it was inversely correlated to empirical reality. Stated somewhat differently, mathematicians had differing opinions about the desirability of the putative separation between mathematics and the physical world, but they all subscribed to the idea that aesthetic considerations came to the fore when empirical content receded. Von Neumann, for example, explained, "As a mathematical discipline travels far from its empirical source, or still more, if it is a second and third generation only indirectly inspired by ideas coming from 'reality,' . . . It becomes more and more purely aestheticizing, more and more purely *l'art pour l'art*."[82] Von Neumann, like Lax, believed that such a situation should be avoided, and argued that mathematics should never lose its attachment to objective reality. However, he also believed that the aesthetic quality of mathematics was inversely correlated with its bond to reality.

It is worth pausing to note that this opinion was far from obvious. The relation between mathematics and art had historically centered on geometry. Even David Hilbert, who developed axiomatics, held, according to Leo Corry, that it was the intuitive rather than the axiomatic or analytic approach to geometry that had "aesthetic and pedagogical" value.[83] It was the aesthetic dimension that Hilbert sought to disseminate in *Geometry and the Imagination*. As Hilbert writes in the preface, "It is our purpose to give a presentation of geometry, as it stands today, in its visual, intuitive aspects." Approaching geometry "through figures that may be looked at" instead of formulas, Hilbert explains, will contribute "to a more just appreciation of mathematics."[84] Aesthetics in mathematics had long been linked to visible form.[85]

This is not to say that midcentury mathematicians were the first to link mathematical aesthetics with nonvisual concepts. Hilbert's contemporary Henri Poincaré explained in *Science and Method* that scientists study nature not because it is useful but because it is "beautiful" and they take pleasure in it. He wrote, "I am not speaking, of course, of that beauty which strikes the senses, of the beauty of qualities and appearances. I am far from despising this, but it has nothing to do with science. What I mean is that more intimate beauty which comes from the harmonious order of its parts, and which a pure intelligence can grasp."[86] For Poincaré, simplicity and unity were constitutive of the aesthetic in science. Both Morse and von Neumann built on Poincaré's ideal of beauty when they discussed the aesthetic dimension of mathematics. Moreover, these ideas were in no way limited to mathematicians. Many other scientists believed that attributes such as simplicity, elegance, and uniformity constituted an aesthetic dimension of their fields more broadly. But this is key: the fact that aesthetic considerations could be and were attributed to physical theories only serves to underline midcentury mathematicians' assertion that a mathematical aesthetic was inversely correlated to reality.

Von Neumann, for example, settled on aesthetics as the element that, by definition, separated the work of mathematicians from that of other scientists. Comparing mathematicians to theoretical physicists, von Neumann explained that while the latter were also concerned with aesthetics, their work was guided from "outside." Not so for mathematicians, whose criteria for both selecting a problem and judging its success were "mainly aesthetical."[87] Von Neumann collected his ideas in a talk he delivered in 1946 as part of a series of lectures at the University of Chicago dedicated to the creative process in the arts and scholarship. The lectures, which were later published as *The Works of the Mind*, also included talks by Marc Chagall on "The Artist," Arnold Schoenberg on "The Musician," and Robert M. Hutchins (then president of the University of Chicago) on "The Administrator." In his talk on "The Mathematician," von Neumann explained that mathematicians sought simplified theories that could join and explain previously special cases, but in such a theory they also expected "'elegance' in its 'architectural,' structural makeup." A theorem, he continued, should not be lengthy and complicated for no apparent reason, and there should always be a clear general principle. "These criteria," he said, "are clearly those of any creative art, and the existence of some underlying empirical, worldly motif in the background—often in a very remote background—overgrown by aestheticizing developments and followed into a multitude of labyrinthine variants—all this is much more akin to the atmosphere of art pure and simple than to that of the empirical sciences."[88] Purity was a correlate of beauty.

By the 1960s, the inverse relation between mathematical aesthetics and empirical reality came to define mathematics tout court. In 1966, statistician

John Tukey, trying to account for mathematical unity, asked whether it still made sense to consider mathematics as a singular entity. Following the work of his colleagues, Tukey said of symbolic reasoning, "In the inner citadel both chains of symbolic reasoning and the results reached by these chains are mainly *judged by aesthetic* and intellectual standards of beauty, universality, economy. At the outer fringes, although beauty, universality, and economy are still valued, progress toward the empirically verifiable, especially toward the prediction or control of events in the real world of objects and men, becomes the prime criterion."[89] To a certain degree, mathematicians formulated a negative definition of aesthetics as the elimination of any empirical content. In this sense, their emphasis on aesthetic autonomy resonated with a broader midcentury ideal most closely associated with art critic Clement Greenberg.

Greenberg had a profound influence on postwar American art criticism. He was an early supporter of Abstract Expressionism, championing the works of Jackson Pollock, and he later defended post-painterly abstraction. In 1939, Greenberg began formulating his theory of formalism by arguing that artists inaugurated modernism in art when they rejected external influences on their work, turning instead toward self-reflexivity. In his 1940 essay "Towards a Newer Laocoön" Greenberg wrote, "As the first most important item upon its agenda, the avant-garde saw the necessity of an escape from ideas, which were infecting the arts with the ideological struggles of society. Ideas came to mean subject matter in general."[90] The rise of artistic abstraction, the turn away from representation, and the rejection of direct "subject matter" were all, in Greenberg's view, a withdrawal from the "ideological struggles of society." In the process, he argued, artists began emphasizing form over subject, and the arts emerged as "absolutely autonomous, and entitled to respect for their own sakes, and not merely as vessels of communication."[91] For Greenberg, abstraction effectively separated aesthetics from social and political concerns.

Greenberg theorized not only modernism but also art criticism itself. As he explained, the autonomy of art was not achieved simply by evacuating it of subject matter, but also by turning toward form. This shift is perhaps most evident in his promotion of medium specificity. As Greenberg explained, "A modernist work of art must try, in principle, to avoid dependence upon any order of experience not given in the most essentially construed nature of its medium. This means, among other things, renouncing illusion and explicitness. The arts are to achieve concreteness, "purity," by acting solely in terms of their separate and irreducible selves."[92] Paintings, Greenberg wrote, were defined by the two-dimensional flat canvas.

Representational art should be avoided not because it was inherently problematic but because it entailed the illusion of three-dimensionality. Such self-reflexivity was a bedrock of Greenberg's overarching theory of aesthetic autonomy.

It might seem odd to place axiomatics, a mathematical technique, alongside Greenbergian formalism. There is nothing in axiomatic thinking that turns an art critic's attention to the compositional elements of a painting, the lines, edges, or flatness of the canvas. And yet, axiomatics and formalism do share a similar genealogy and certain characteristics. Greenberg's formalism was influenced by early-twentieth-century Russian Formalism, which brought together linguistic and literary theory and placed an emphasis on syntax over semantics. Lynn Gamwell has drawn an even closer connection between Hilbert's axiomatics and Russian Formalism, explaining that "sensing a common purpose with formalist mathematicians (as well as linguists), Russian Constructivist artists and poets in Moscow, Saint Petersburg, and Kazan reduced their visual and verbal vocabularies to meaning-free signs and composed them as autonomous structures."[93] Both formalism and axiomatics emerged as a response to a crisis of referentiality, and while distinct, axiomatics and formalism responded to the problems of signification and referentiality by emphasizing structure over content.[94]

Another genealogy of midcentury formalism can be traced to the work of the British art critics Roger Fry and Clive Bell on the one hand and New Criticism on the other. Fry and Bell were members of the Bloomsbury Group, and were well acquainted with Russell and Whitehead's work.[95] As Gamwell points out, it is appropriate that Fry's art criticism came to be known as formalism:

> Just like Hilbert, Frege, and Russell all arrived (by different routes) at the same basic idea of an axiomatic structure, similarly Bely, Shklovskii, and Fry all (for different reasons) ended up emphasizing form over content. In resonance with Hilbert's focus on mathematics as a formal axiomatic structure, the Russian/German formalist artists and poets composed with shapes and sounds that do not represent the natural world-out-there. Inspired by Frege/Russell's focus on logic and spoken language, British formalist artists used forms to depict ("picture") the natural world, as seen in the figures and landscapes of the Bloomsbury Group.[96]

What perhaps distinguished Greenberg's formalism and aligned him perfectly with postwar pure mathematicians was that he pushed the practice to its extreme. Fredric Jameson points to Greenberg as the theoretician who, more than any other, helped to propagate the "ideology of modernism," which he defined by its nearly exclusive commitment to the autonomy of the aesthetic. Jameson argues that this ideology was an "American invention" and "a product of the Cold War."[97] It was not constitutive of modernism itself. In the aftermath of World War II, Jameson explains, the West (and the Stalinist East) desired a "stabilization of the existing systems" and an end to the upheavals that had been enacted in the name of modernism before the war. In this new environment, "social impulses" were to be "repressed or disciplined."[98] Even if they appeared on the surface to operate under the shared banner of modernism, the social vision of interwar artists, Jameson argues, was different from that of postwar abstract artists. In calling the ideology of modernism a Cold War invention, Jameson sought not to dismiss the works of Abstract Expressionism but rather to draw attention to the *theory* of art that developed alongside it.

"Greenberg's greatness as an ideologist," writes Jameson, was not only in having offered aesthetic autonomy as the culmination of the modernist movement. He also "grasped the onset of the Cold War not as the end of hope and the paralysis of the productive energies of the preceding period, but rather as the signal opportunity to forge a brand-new ideology that co-opts and re-awakens those energies and offers a whole new (aesthetic) blueprint for the future."[99] Once steeped in Marxist ideology, Greenberg had, by the end of World War II, posited *both* an abstract art and a formalist criticism that was by definition apolitical. As he wrote in "The Case for Abstract Art," abstract art was significant because it was the necessary precondition for disinterested contemplation: "Abstract art comes, on this level, as a relief, an arch-example of something that does not have to mean, or be useful for, anything other than itself."[100] Like New Critics, Greenberg positioned abstract art in opposition to the ascendency of technoscientific knowledge. Sounding remarkably similar to Morse and his colleagues, Greenberg wrote, "If American society is indeed given over as no other society has been to purposeful activity and material production, then it is right that it should be reminded, in extreme terms, of the essential nature of disinterested activity."[101] The dual nature of abstraction is thus also evident in Greenberg's thinking: Abstraction, he suggested, was both an inherent characteristic of the work itself and, as a verb—to *abstract*, to *withdraw*—a condition of its production. Aesthetic autonomy, for mathematicians and artists, was a concomitant of this dualism.

With the hindsight afforded by the 1980s, Peter Lax similarly identified politics as the suture between midcentury mathematics and midcentury art: "Next to Bourbaki, the greatest champions of abstraction in mathematics came from the American community. This predilection for the abstract might very well have been a rebellion against the great tradition in the United States for the practical and pragmatic, the postwar vogue for Abstract Expressionism was another such rebellion."[102] Mathematicians and artists were coming to terms with the legacy of abstraction in an ideological and political reality shaped by the Cold War. Remarkably, despite the obvious and profound differences between the two communities, their struggles with abstraction—its meaning, its justification, and its place in postwar America--paralleled one another.

L'art pour l'art did not emerge with Greenberg, nor did the idea of pure mathematics as an autonomous intellectual pursuit first appear with midcentury mathematicians. The Romantic ideal of art could just as easily be applied to representational art, and the idea of mathematics as the intellectual ideal par excellence emerged before the twentieth century. But it was only at midcentury that abstraction came to hold this double meaning. In mathematics, this duality depended on the proliferation of modern axiomatics; in the arts, it relied on formalism. The two—axiomatics and formalism—helped create the fertile ground in which abstraction proliferated in Cold War America.

A Useful Contradiction

Mathematicians based their claims to artistic conception in modern axiomatics. In turning away from the world and, more importantly, elevating axiomatics to a full-fledged research methodology, mathematicians refashioned themselves as artists. They argued that modern axiomatics had bestowed upon them the freedom necessary to express their creativity and pursue their work as an artistic activity. In his 1943 *Science* article, Morse pointed to this notion when he explained that the imagination afforded to mathematicians was achieved partially by the ability to "have that freedom that comes from recognition of dogma, and the open acceptance or rejection of an axiom at will."[103] Since the axioms of a theory were not determined by empirical reality and were highly arbitrary, settling on an axiomatic system was in itself a creative act. Modern axiomatics gave credence to mathematicians' claim that mathematical theory was unbound from the world. It offered a direct methodology by which an otherwise philosophical claim turned into a clear research program.

The analogy between axiomatic research and artistry is best illustrated in a national report presented to the National Academy of Sciences in 1954:

> The axiomatic approach has emancipated mathematics from its bound state in science . . . At the same time, it develops a deeply artistic aspect in its own nature. The structures which the mathematician axiomatizes are, in final analysis, his to choose and to change, modifying, dropping, or adding an axiom here and there—much as a score evolves under the composer's hand. There is no longer a need to scan the shifting reflections in the pool of his mind for the features of an alien reality forever looking over his shoulder. Instead, there unfolds a new and richly fascinating world.[104]

Axiomatics transformed mathematics into modern art. The freedom afforded by the axiomatic method guided mathematicians in this new terrain.

In a 1961 lecture on "Characteristic Features of Mathematical Thought," German mathematician Johannes Weissinger similarly identified axiomatics as the defining feature of modern mathematics. By eliminating intuitive evidence from mathematical research, he argued, mathematicians achieved a higher degree of certainty and a new level of freedom: "In free artistic play, guided only by a sense for mathematical values, one can modify, omit, and add individual axioms—or for that matter, create and examine entirely new systems of axioms."[105] Axiomatics defined modern mathematics and became the building blocks of mathematical creativity.

Here is a good place to pause. Given the previous two chapters, it should come as a surprise that mathematicians now credited axiomatics for rendering their work a creative and artistic pursuit. After all, as I have shown, mathematicians also pointed to modern axiomatics to account for the growing applicability and utility of mathematics across all fields of knowledge. Indeed, the above quotation from the national report appeared in a survey of *applied*, not pure, mathematics; in the following pages, axiomatics is closely linked to mathematical applications. Mathematicians, however, saw no apparent contradiction in claiming that axiomatics was responsible for both the utilitarian and nonutilitarian aspects of the discipline. Axiomatics stood for a break between theories and physical reality. Pure mathematicians understood this severing in terms of the freedom it provided them to pursue knowledge for its own sake, not motivated by external concerns. Applied mathematicians, on the other hand, conceived of the separation as denoting a form of analytic realism. Mathematics was a useful tool precisely because it provided a method with which to gain deep knowledge of what lay beneath the visible surface. This contradictory nature of abstractionist theory helped it to become so pervasive in Cold War America. Mathematics thus offers a clue to Cold War thought more broadly.

By centering on abstraction, postwar social sciences and formalist art come into relief not as two diametrically opposed tendencies, but rather as *two sides of the same coin*. Social scientists turned to mathematics during the postwar period as a tool of objectivity and a way to align their work with the hard sciences. As Theodore Porter writes, "For the Cold War generation, disciplinary autonomy and the assertion of neutral objectivity were not merely desiderata of social science but defining characteristics."[106] Porter sees the social sciences' turn toward scientism and neutral objectivity as partially a reaction to anti-Communist sympathies and McCarthyism. It was a way of depoliticizing knowledge production while emphasizing "independence and detachment."[107] However, as Porter writes, social scientists "depoliticized" knowledge not to separate their work from the social world but, paradoxically, in hopes of making their work more relevant, to position it as the basis for solving social problems.

It was not mathematics per se that enabled social scientists to achieve this position; rather, it was axiomatic thinking. Social scientists could have continued to rely on mathematics for measurement and calculation, as they had in the prewar period. Instead, abstract axiomatic theories offered them a means of producing presumptively depoliticized knowledge that was nonetheless socially useful (or at least claimed to be). Abstraction enabled social scientists to claim utility and neutrality at the same time. Greenberg and other formalist critics positioned themselves in direct opposition to the bureaucratic objectivity promoted by postwar social scientists, but they, too, claimed their work as depoliticized.

In advancing a formalist reading of literature, New Critics called upon an ideal of objectivity. Allen Tate explained, for example, "The formal qualities of the poem are the focus of the specifically critical judgement because they partake of an *objectivity* that the subject matter, abstracted from the form, wholly lacks."[108] Stripping away external considerations allowed New Critics to elevate formal criticism to the status of disinterested inquiry. New Critics' appeal to objectivity garnered derision as nothing more than a bad case of science envy. As Gerald Graff shows, "The New

Critics' attempt to emulate the empirical scientists by applying an objective, analytic method to literature was judged as one more symptom of the university's capitulation to the capitalist-military-industrial-technological complex."[109] However, as Graff convincingly demonstrates, this view mistakes the technique of close reading with a theory of literature. Far from capitulating to the empirical sciences, New Critics envisaged literary criticism as an antidote to positivistic science.

The same holds true for Greenberg. As Caroline Jones argues, Greenberg's formalism "became an analytic, subtly reconfigured . . . to justify a particularly positivist view of abstract art."[110] Formalism, like New Criticism, "refused to 'interpret,' rejecting biography, intention, and symbolism in favor of concrete material 'facts.'"[111] Moving away from content, art criticism was charged with paying attention to things like the surface of pigments, the dimension of the canvas, the relations between edges, colors, and lines. Humanists did not find the ideal of objectivity suspect in itself, but rather took issue with its being co-opted in service of the bureaucratization of daily life. Humanist critics nevertheless called upon objectivity to frame their work as nonideological. As Yve-Alain Bois explained, "Greenberg pretended to a kind of ideological neutrality; this allowed him to make his positivist turn of seeming simply to describe the art . . . Everything was discarded: an ex-Marxist, Greenberg could present himself as objective—as objective as an engineer or a scientist. So he became the pure empiricist who only describes."[112]

Formalist critics' positivism was, of course, not built on mathematical axiomatics, but it shared the abstractionist tenets that emphasized structure over content and analytic attention to relations.[113] However, their goal was not to make art more useful or applicable. Like pure mathematicians, they embraced abstraction as a technique that allowed them to elevate their work above the political and social morasses of the day.

Abstraction was so forceful in Cold War America because it symbolized the ideology of no ideology. Both Porter and Jones acknowledge that the claim to ideological neutrality was itself a postwar American ideology. Porter writes that social scientists' "preoccupation with neutral objectivity can itself be seen as a form of politicization by virtue of its very claim to stand outside the value-laden character of the processes and interests that shaped the production and uses of social knowledge."[114] Jones, for her part, notes that "Greenberg was willing to use formalism to inculcate art against ideology (an American postwar ideology unto itself)."[115] And mathematicians paved the way.

Pure mathematicians' claim to the arts was more than just rhetoric. As I demonstrate in chapter 5, it had real implications for the institutional development of the field in the postwar period. It enabled pure mathematics to grow and expand while the development of applied mathematics remained stalled.

CHAPTER FIVE

Unreasonable Abstraction

The Meaning of Applicability, or the Miseducation of the Applied Mathematician

The paradox is now fully established that the utmost abstractions are the true weapons with which to control our thought of concrete fact.
—Alfred North Whitehead

In 1971, University of Wisconsin mathematician Richard C. Buck sent a circular discussion paper to several of his mathematical colleagues urging them to consider the "argument for federal support for mathematics." Two years earlier, Buck, who was then in his fifties, had helped mobilize the mathematical community in support for mathematical students whose studies had been abruptly stopped either because they were serving time in prison for refusing the draft or were involuntarily serving in the military. Together with Chandler Davis and John L. Kelley, both of whom were dismissed from the University of California, Berkley, in the 1950s for refusing to sign the loyalty oath (Kelley) and cooperate with the House Un-American Activities Committee (Davis), Buck collected a list of leading mathematicians who offered to volunteer their time and help the impacted mathematicians continue their studies remotely. In 1971, the Vietnam War was still raging, and Buck was once again concerned about the welfare of young mathematicians, but much had changed in the intervening two years. Buck was now concerned not with a fraction of mathematical students, but with a full cohort of PhDs who were facing unemployment.

Buck's memorandum came at a time of great upheaval for the mathematical (as well as scientific) community. The post–World War II era was coming to an end, and almost overnight, scientists began experiencing what Daniel Kevles describes as "an employment squeeze reminiscent of the 1930s depression."[1] Historians have noted that several factors contributed to this transformation. A rise in expenditures forced universities around the nation to tighten their budgets, while these universities concomitantly lost the endorsement of the general public. Research for its own sake was no longer deemed a defensible cause. Instead, an emphasis on applicability and public service was placed front and center.[2] As far as scientific research was concerned, however, it was the overall decrease in federal support for research and development and the changing emphasis on applied as opposed to basic research that had the greatest effect.

When universities began cutting down their faculties and the National Science Foundation (NSF) decreased its funds earmarked for mathematical research and education, the mathematical community, together with the rest of the sciences, absorbed much of the shock. Unemployment began to rise, and anxiety regarding the future of the field became an unavoidable reality. It was with these concerns in mind that Buck urged his colleagues to reconsider what might be the argument for federal support for mathematics: "We can resort to polemic, point with pride, and raise the spectre of competition with the USSR and Europe . . . This has some psychological strength, but it is certainly weak in logic."[3] Unlike other scientific fields, mathematics, Buck explained, was not externally generated. Thus, the only argument for federal support for mathematics was based on the notion of applicability. However, he added, "It seems hypocritical . . . to sell it [mathematics] by a claim of applicability, unless we are willing to give the latter full justice in training programs for mathematicians."[4]

As the unemployment crisis continued throughout the early 1970s, many other mathematicians pointed to the insular nature of graduate education in the field as a cause for concern. This was a moment of reckoning for the mathematical community as a whole.[5] Much ink has been spilled assessing the influence of the growing national security state on intellectual activities in the United States during the postwar period. Looking to physics, oceanography, and area studies, scholars have asked to what degree military demands and military funding directed intellectual production in and outside the academy.[6] It seems that no academic field was left untouched by the growth of what Stuart Leslie has called the military-industrial-academic complex.[7] As Naomi Oreskes has pointedly put it, "What difference does it make who pays for science? The short answer is: a lot."[8]

It is enough to look at the growth of computing, game theory, mathematical statistics, and fluid dynamics in this period to demonstrate how implicated postwar mathematical theory was in specific military and industrial concerns.[9] The growth of these fields was dependent on federal funding and military interests. However, it is harder to account for the fact that during the same period, pure mathematics became more abstract and divorced from the lived world.[10] And when it came to graduate education in mathematics, it was *pure*, not applied, mathematics that represented the lion's share of graduating PhDs in the field. Moreover, those students who took their degree in pure mathematics were more isolated in their studies than any generation before. Many doctoral programs in mathematics did not require any advanced training in adjunct scientific fields, making graduate students even less suitable for nonacademic employment.[11]

Within traditional academic departments, the postwar growth of the field was defined by pure mathematics with its high modernist emphasis on abstraction and the autonomous conception of the field. This is why mathematicians like Buck immediately turned to graduate training when the funding began to disappear. If mathematicians were going to argue that their research and educational activities should be supported by the federal government because of the applicability of mathematics, then the least they could do, he argued, was to emphasize this point in their graduate programs.

The growth of academic applied mathematics in the postwar period challenges some of the basic scholarly assumptions about the growth of science during the Cold War, most specifically the idea that researchers oriented their scholarship to fit the requirements of funding agencies. Despite the obvious interests of the defense industry in applied mathematics (fields such as aerodynamics, mechanics, cryptography, etc.), within traditional academic departments, the field remained stalled. In the 1950s, most applied mathematical research was conducted in newly established semiautonomous research institutions that were almost fully dependent on governmental contracts for their existence. The Applied Mathematics and Statistics Laboratory at Stanford, the Institute of Fluid Dynamics and Applied Mathematics at the University of Maryland, the Institute of Mathematical Sciences at New York University, the Graduate Institute for Mathematics and Mechanics at Indiana University, and the Institute for Numerical Analysis at the University of California, Los Angeles, were all postwar products. "Research and advanced training in applied mathematics," explained a 1954 report, "is at present a ward of the Federal Government."[12] Military funding helped spur research in the field, but it also *hindered* its development by closely aligning its growth with utilitarian demands and setting it apart from academic pure mathematics.[13]

Oddly enough, when, in the aftermath of Sputnik and the passage of the National Defense Education Act in 1958, support for mathematics research and education mushroomed, research and education in applied mathematics remained marginalized in universities. During the 1960s, the number of annual mathematics PhDs per year quadrupled from 303 in 1960 to 1,204 in 1970. Throughout the decade only 15 percent of these new PhDs were in applied mathematics with an average of 65 percent in pure mathematics (the rest were in statistics). Moreover, computing, statistics, and operations research, which at first were taken to be part of applied mathematics, grew into distinct disciplines independent of mathematics, and applied mathematics continued to falter.[14] When it came to applied mathematics, funding was not enough. Mathematics, at least in its academic guise, seems to defy historical wisdom.

Mathematicians were able to maintain a high degree of autonomy by homing in on abstraction. First, they pointed to the processual nature of mathematical abstraction to argue that pure mathematics was closely related to applied mathematics while being *simultaneously* independent of it. Second, they maintained that it was exactly the abstract nature of pure mathematics that gave rise to its increasing applicability, or what they called the "mathematization of culture."[15] As demonstrated in the previous chapters, both applied and pure mathematicians believed that modern axiomatics was responsible for the increasing applicability of mathematics. However, this argument on its own could not ensure that pure mathematical research would be eligible for federal support. To argue that research in category theory had potential future utility, mathematicians had to reclaim abstraction as utility. Their success guaranteed that mathematicians enjoyed the fiscal benefits of the Cold War like the rest of the sciences while maintaining incredible autonomy over the development of their field. Thus, while scientists and some social scientists were leaving the ivory tower, mathematicians continued to center their activities in and around the university, resembling most closely other humanistic fields.[16]

As a consequence, the growth of applied mathematics as an independent academic field was stymied during the postwar period. As Buck noted in his 1971 letter, the case of the applicability of mathematics was "hypocritical" not because mathematics was not applicable, but because mathematicians did not try to train their students with this belief in mind. Historians of science have long challenged the idea that a clear demarcation exists between basic and applied research, as well as the notion that innovation only flows in one direction from the basic to the applied.[17] As Oreskes has shown, at times it has been the most applied problems that have led to new theoretical understanding. However, the growth of pure and applied mathematics after the war challenged the very nature of scientific utility and how it was constructed and maintained.

On the Nature of Utility

Mina Rees developed a love of abstract algebra as an undergraduate student at Hunter College. Upon graduating, she enrolled at Columbia University, but after two years of study it was clear to her that "the Columbia mathematics department was really not interested in having women candidates for Ph.D.s."[18] This, she recalled later, was the first time it occurred to her that some members of the mathematical community did not believe it was an appropriate study for women. She transferred to Columbia University's Teachers College and earned a master's degree in education. She was not, however, one to give up easily. In 1929 she arrived at the University of Chicago to pursue a PhD, and a year later during the summer meeting of the American Mathematical Society, she had breakfast with both Marston Morse and George Birkhoff. "I was enchanted with mathematicians and, at least partially because of that, with mathematics," she said.[19] PhD in hand, she returned to New York City and began teaching at Hunter College, which was the only possible career path at the time for a woman with a mathematics PhD.[20]

Her mathematical career took a sharp turn during World War II, when Warren Weaver appointed her first as his technical aid and later as executive assistant at the Applied Mathematics Panel. During the war she met many members of the mathematical community and gained firsthand knowledge of the extent of applied mathematics research in the US. In the war's aftermath, she was chosen to be the first head of the Mathematical Section of the Office of Naval Research (ONR). Before the establishment of the National Science Foundation in 1950, the ONR was the first funded organization dedicated to supporting basic scientific research and Rees was charged with directing its expenditures to mathematics. First, though, she had to find a place to live, which was no easy task in 1946 when she moved to Washington. There were no available apartments when she got there, and most hotels had a five-night maximum stay. When she found a hotel that allowed guests to stay for two weeks, she hit upon a solution. Every two weeks, she would vacate her room and travel to a different mathematics department around the country. When she returned, she registered again for two weeks.[21] During her time at the ONR, Rees helped sketch out and implement postwar support for mathematics.[22]

In January 1948, Rees published an article in the *Bulletin of the American Mathematical Society* in which she sought to explain to members of the society "the philosophy which ha[d] determined the mathematical research project which ONR [was] sponsoring."[23] Rees explained that as a matter of policy it was recognized early on that in order to best support mathematical research in the United States, research in pure mathematics must also be included among the branch's contracts. Yet, noting the dearth of research in applied mathematics at the start of World War II, Rees established that the lion's share of navy support went to applied research: "It is a fact that over 4/5 of the annual mathematical expenditure is in support of research in 'applied mathematics,' mathematical statistics, numerical analysis and computing devices."[24] The ONR deemed these fields, Rees explained, to have the highest priority given their potential future applicability to the navy's affairs.

Six years later, she announced that the ONR had decided to cut back its program in pure mathematics. Underlying this change in policy, Rees explained, was the establishment of the NSF, whose sole purpose was to support basic research in the sciences. Writing in *Science*, she explained, "It is natural that, in mathematics, it [the NSF] should support an outstanding program in some of the more abstract fields, where much of the most significant research is going on."[25] Rees was trying to assuage the fear of many in the mathematical community, reassuring them that despite the change in overall support to mathematics, research would not decrease, but her efforts at first were unsuccessful.

What mathematicians realized was that, at least initially, the establishment of the NSF did not solve the problem of support when it came to pure and applied mathematics. In January 1953, William Duren, who had just completed a short term as acting program director for Mathematics at the NSF, published an article in the *Bulletin of the American Mathematical Society* regarding support for mathematical research. His experience, Duren maintained, had made one thing clear: the NSF Act of 1950 did not "in itself create a new era in which a non-military arm of the Government [would] support *basic* research in theoretical mathematics without the demand for direct military or even physical science relatedness."[26]

Mathematicians, Duren noted, did not seem to realize that research in mathematics still needed to be justified if it was to receive any support.

Like the social sciences, support for pure mathematicians was not guaranteed at first when the NSF was established. As Mark Solovey has shown, some natural scientists and administrators argued that the social sciences should be left out of the NSF for not being scientific enough and for their association with left-wing ideology. To gain access to the funding opportunities presented by the NSF, several social scientists began insisting that their work was scientific and had a social utility, by advancing American progress at home and abroad. Pure mathematicians did not have to argue for the objective nature of their work. Perhaps more than any other science, mathematical theory was seen as value-neutral. However, they did have to account for its utility. And, while the social scientists did so by directly considering "national needs and the type of knowledge that could best advance the healthy development of the United States as a dynamic modern society and leader of the free world," mathematicians did so by appealing to abstraction.[27]

Chicago mathematician Adrian Albert was one of several pure mathematicians who in the 1950s forcefully advocated for support for pure mathematics. During the war, Albert worked for the Applied Mathematics Panel and also became interested in cryptography, but his main area of research was abstract algebra. In 1952, as chair of the Division of Mathematics at the National Research Council, Albert wrote a memorandum on the support of mathematics to the members of the Division of Mathematical, Physical, and Engineering Sciences of the NSF. Albert explained that the reason for the memorandum was the mathematical community's strong dissatisfaction with support for mathematical research. "The fundamental role of mathematics in modern science," Albert began, "becomes increasingly evident daily as mathematical concepts, techniques, and *modes of thought* are adopted in sciences from physics to the study of social behavior."[28] However, this importance was not reflected in the support the field had received so far. It was the goal of his memorandum, he emphasized, to increase the NSF's overall support of the field.

To make his case, Albert listed the subfields of mathematics, which according to him were algebra, analysis, geometry, and topology. "The field called applied mathematics," he wrote, "has not earned the right to be called a major field since the great advances in mathematics have taken place in the pure fields listed above, and it is the ideas, techniques, and modes of thought of these fields which are finding the broad applications to other sciences today."[29] According to Albert, applied mathematics, at least as developed in the United States, was just not sufficiently advanced to warrant equal representation among the rest of the mathematical fields.[30]

Albert reported that despite these facts, recent support patterns of the NSF revealed that 77 percent of the budget afforded to mathematics went to fields of applied mathematics, 17 percent to analysis, and only 6 percent to other fields of pure mathematics. Given the significance of research in pure mathematics, Albert suggested, there was a clear imbalance in the foundation's allocation of funds to the field. This meager share for support in pure mathematics was even more worrisome, Albert argued, given that research in applied mathematics, unlike studies in pure mathematics, was already receiving support from military bodies. There was no reason, according to Albert, that four out of the eight mathematics consultants for the National Science Foundation were applied mathematicians. Albert elaborated on that point in an earlier draft of the report. The field of applied mathematics, he wrote, could "be readily justified as a 'related' topic by a military agency," and so its support [would] undoubtedly continue. Moreover, since analysis served as the bedrock of most work in the field, "the concept of *relatedness* [could] be stretched to include that field also."[31] Despite the fact that "the really fundamental advances in mathematics" had been in "the more abstract parts" of the field, they had for the most part been neglected by the various military agencies. According to Albert, mathematicians did not object to this policy by the military, but they nonetheless expected the NSF to balance this tendency by emphasizing its support for pure mathematics.

What Albert, Duren, and many others recognized was that mathematicians needed to reconfigure the relation between pure and applied mathematics according to the prevailing discourse of the time. Published in 1945, Vannevar Bush's *Science: The Endless Frontier* functioned as a blueprint for postwar federal funding for the sciences. Bush's great accomplishment after the war was to convince the federal government to support basic scientific research and as a result, the language of basic and applied research dominated discussions of scientific research. As Bush wrote in his report, "New products and processes [were] not born full-grown," but resulted from "basic scientific research. Basic scientific research is scientific capital."[32] Government support was always understood in terms of future utility. But how to distinguish "basic" from "applied" mathematics? What would such a distinction even mean? The problem was that pure and applied mathematics did not fit neatly within the basic and applied dichotomy that dominated scientific discourse of the period. Applied mathematics did not necessarily equal "applied research." This could be the case, but it was in no way necessarily so. Indeed, until World War II, applied mathematics was almost by definition *basic* research. Research in fluid dynamics or elasticity theory, which belonged squarely within applied mathematics, could be directed toward practical ends, but it could just as easily be labeled "basic" research.

This issue had already created tension during the war. In a November 1943 diary notice following a trip to Washington, Richard Courant reported that recent memoranda issued by the Applied Mathematics Panel (AMP) had been criticized by officers in the Naval Ordnance for being "too academic."[33] Two months later, following a discussion with mathematician Hermann Weyl, Courant wrote to Weaver, "Weyl's papers on shock waves were discussed. They seem to be of such purely mathematical almost axiomatic character that publishing them as N.D.R.C. reports or memoranda might seem objectionable."[34] Even though shock wave theory was of direct interest to the military, and despite the work taking place under the auspices of the AMP, it was nevertheless considered too theoretical to be deemed practical. Weyl's papers could be classified as applied mathematics, but they were basic, not applied research.

When attached to mathematics, that is, the adjective "applied" had become misleading. Unlike mathematization, which denoted the appropriation of mathematics by other disciplines, "applied mathematics" was concerned with the construction of *mathematical theories*.[35] In that regard, the field was more akin to mathematical physics. However, if research in applied mathematics occupied the position of basic research, what position in the new funding landscape would be left for research in pure mathematics? Which subfields of pure mathematics should the NSF support? Did research in algebraic topology deserve to be labeled "basic" research? It was clear that aerodynamic theory had military applications, but what about homological algebra? Who should decide? And how could one assess the utility of those "modes of thought" attributed to pure mathematics?

In the first decade after World War II, the mathematical community could not settle on clear answers to these questions. Rather, grappling with them within the new institutional framework of the postwar period gave rise to increasing tensions between pure and applied mathematicians. In December 1954, Albert reported in an internal university memo that he had just returned from the International Congress of Mathematicians in Amsterdam, where he served as chair of the United States delegation. "During the meetings," he added, "I took the vital steps necessary to settle a very nasty dispute between M. H. Stone, representing the interests of pure mathematicians, and F. J. Weyl, J. Tukey, and S. S. Wilks, representing the interests of applied mathematics."[36] The "nasty dispute" Albert referred to had begun a few months earlier, when a draft of a new national report on the state of applied mathematics in the United States was circulated among leaders in the community. The controversy illustrates the mathematical community's first attempts to come to terms with the new, postwar national funding structure.

Secular vs. Monastic

"If the National Research Council is to support an attack on pure mathematics," Saunders Mac Lane wrote to Leon Cohen in July 1954, "this attack should be specifically labeled as such and not mixed in with a careful analysis of the nature of the applications of mathematics."[37] Mac Lane had just completed reading an early draft of the final report of the Committee on Research and Training in Applied Mathematics and he was angry. A few weeks before he received Mac Lane's letter, Cohen, who only a year earlier had become the program director for mathematical sciences at the NSF, had expressed his own dissatisfaction with an early draft of the report. In a letter to Joachim Weyl, the author of the report, he wrote, "I hope that it will undergo a fundamental revision before publication."[38] In May, a copy of the report was presented at the annual meeting of the Division of Mathematics at the National Research Council, leaving several attendees incensed. Most of the outrage was triggered by a specific section of the report titled "Applied Mathematics in the Scientific Community." Noting that the section had "been interpreted as a violent attack on the American Mathematical Society,"[39] Albert, the current chair of the division, circulated a copy of it among members of the division and leaders in the society.

In response, Marston Morse announced that he would vote for the report only if the "paragraphs with the political implications [were] eliminated."[40] Stone, who was an official member of the committee, was even more outraged. In June, he wrote to Detlev Bronk, president of the National Academy of Sciences, and William Rubey, chair of the NRC, in protest. Weyl and other members of the committee, Stone proclaimed, desired to publish as part of the final report "statements about the mathematicians and mathematical organizations of the country, which seem[ed] to be misleading, offensive, and destructive."[41] He also announced that because attempts to convince the committee to change its report had proven futile, he would file a minority report.

The Committee on Research and Training in Applied Mathematics had been established two years earlier. In April 1952, as chair of the Division of Mathematics, Morse sent a letter to Alan Waterman at the NSF, Mina Rees at the Office of Naval Research (ONR), and representatives of the Office of Ordnance Research and Air Research and Development Command announcing the appointment of the new committee. Its goal, Morse explained, was to study what universities, the government, and industry could do to support research and training in the field. Asking for financial support for the committee's work, Morse noted that the above organizations had a "natural interest" in the results of such a study.[42] Despite the obvious growth of applied mathematics in the aftermath of the war, by the early 1950s the field was not prospering as some had hoped. The confusion that had surrounded the constitution of the field in the immediate aftermath of the war persisted. Moreover, the field failed to take hold in traditional academic departments, both in terms of faculty appointments and graduate training. The committee was in charge of surveying current trends and making future recommendations.

"Applied Mathematics in the Scientific Community" began by declaring that there existed "a deep-running undercurrent of feeling to the effect that the applied aspects of mathematics and those who ha[d] a concern therewith fail[ed] to receive their due respect and recognition in the representative organizations of American mathematics."[43] It suggested that since the philosophy and interests of these organizations affected the development of the field writ large by influencing students, publications, and the allocation of funds, it was of paramount importance that these organizations had an accurate understanding of the constitution and importance of the field.[44] As noted in chapter 2, in its overall conception of mathematics, the report adopted a similar view to that of pure mathematicians by emphasizing modern axiomatics. The grounds for the dispute, as such, were not intellectual but social.

The report identified three factors that impeded one's true appreciation of applied mathematics. The first misconception the report singled out had to do with the supposed "creative freedom" the applied mathematician, "especially when in non-academic employment, enjoy[ed],—or, rather fail[ed]s to enjoy." Since the applied mathematician was "supposedly bound to produce specific results," he tended to be seen as "a professional craftsman, plying a trade he learned," as opposed to doing original and creative work. The second misimpression the report identified was the belief that applied mathematicians only solved "dull problems by repulsive means." Finally, there was an "impalpable ever-present haze of suspicion" that abstractly minded mathematicians could, if so inclined, do a better job than "their applied cousins by the bothersome problems of the world around us."[45] All three misconceptions, the controversial section suggested, prevented applied mathematicians from fully taking part in American universities. The report emphasized that what hampered the growth of applied mathematics was not a lack of funding or institutional support, but deeply rooted ideas about the supposed hierarchy between pure and applied mathematics.

A main agitator behind the report was Richard Courant. In January 1954, he pleaded with Weyl to emphasize in his writing the sociological aspects impeding the growth of applied mathematics. "The rulers of mathematics as an activity in the academic world," Courant explained, tended to exclude applied aspects of mathematics from influence.[46] He added, "Vital interests, not only of the insignificant mathematical fraternity, but of the country, and I dare say of the free civilization" were at stake. Courant was one of the many German Jewish émigrés who had arrived in the United States during the 1930s. A well-respected mathematician and an expert in mathematical physics, Courant had received his PhD in 1910 under the supervision of Hilbert. Upon his arrival in 1934, Courant accepted an appointment at New York University, where he set out to establish a mathematical school dedicated to the close interaction between pure and applied mathematics. He was joined three years later by two young mathematicians, Kurt O. Friedrichs and James Stoker. Friedrichs had been forced out of Germany after marrying a Jewish woman. Stoker, who was born in Pittsburgh, was recommended to Courant by George Polya, with whom he had spent time in Zurich. Over the next two decades, the three men served as the core of the mathematical group at NYU. They collaborated on numerous projects, wrote papers and books together, and trained an entire generation of American applied mathematicians. Courant supervised twenty-two students after his arrival at NYU, Friedrichs surpassed him by training thirty-six, and Stoker managed to exceed Friedrichs by one. Yet their road was rocky and often felt as though it ran uphill.[47]

Not long after his arrival in the United States, Courant began warning against what he perceived as the unbalanced emphasis on pure mathematical research in the country. In 1941, Courant published a book entitled *What is Mathematics?* in which he asserted his vision for the future development of mathematics. The most important task facing mathematics in future years, he announced, was "to establish once again an organic union between pure and applied science and a sound balance between abstract generality and colorful individuality."[48] War mobilization was the perfect condition under which Courant could advance his cause. The group Courant gathered around him began consulting on a range of military problems as soon as mobilization kicked into high gear, and once the Applied Mathematics Panel was established, the relationship was formalized. Courant worked closely with Weaver during the war, and the two men shared similar aspirations for applied mathematics. However, this initial reliance on the defense establishment would become a double-edged sword for Courant's group. After the war ended, the group remained completely dependent on military projects to support its operations.

The first contract grant Courant's group signed after the war was with the Office of Naval Research for a study on "fluid dynamics and mathematical physics." This initial contract extended more than a decade and was supplemented over the next couple of years by additional contracts. For example, in 1951 the ONR signed another contract with the group for a study of the "formation of jets and their stability." The ONR was not, however, the only military agency that began supporting the group's research. The Army Department signed a contract in 1952 for a study of the "theory of subsonic and transonic fluid dynamics." This reliance on outside contracts steadily increased in the decade following the war. By 1958 Courant's group was conducting research for twelve separate agencies, from the Air Force and the National Security Agency to the National Science Foundation.[49] Its annual budget rose from approximately $125,000 in 1946 to about $2.5 million in 1958.[50]

Courant's group at NYU was completely dependent on these outside contracts. It was not only used to support the research staff and graduate students, but even some of the salaries of the senior mathematicians in the institute were paid through contract money. This dependency on contract money set applied mathematicians apart from their peers in pure mathematics, who were supported more directly by their universities and were therefore free to pursue research of their own accord. In 1953, commenting on the support structure for the institute, Courant noted,

"We feel that this support can be accepted with good conscience. We are trying to help our sponsors in a number of partly classified specific and important subjects. Before all, we are convinced that our existence and our work contribute to creating a reserve of competent people which would be prepared to help if and when an emergency should occur." Yet he immediately added, "Such one-sided support obviously is not a healthy basis for our far-reaching objectives."[51] In enumerating the future objectives of the institute, Courant explained that despite the fact that "in the present emergency" a fair amount of classified work was justified, the emphasis in the future should be on "*basic* research."[52]

Other applied mathematics institutes that were set up after the war fared similarly. For example, Albert Bowker, who served as the first director of the Applied Mathematics and Statistics Laboratory at Stanford, described the group as "a kind of a holding company for government projects."[53] In addition to the source of funding, the organizational structure of these groups also set them apart from pure mathematics. The Stanford Laboratory established in 1950 was modeled directly on the Applied Mathematics Panel's Statistical Research Group at Columbia University, and some of the early statistical work it produced, on sampling inspection, for example, originated in war research. During the 1950s, the scope of projects expanded to include research in mathematical economics, mathematical techniques in the social sciences, and classical applied mathematics, but the organizational pattern remained the same. Bowker explained, "The idea from the beginning was to construct a research laboratory with students and faculty working on problems, many of which would come from applied fields; to treat students as colleagues."[54] Such emphasis on collaborative research was alien to most pure mathematicians of the time.

Thus, while military and federal funding had been crucial to the growth of applied mathematics, in the war's aftermath it also set applied mathematics apart institutionally, organizationally, and financially. The emphasis on teamwork, organized projects, and outside support created a clear structural distinction between research in applied and pure mathematics. This explains why Courant pushed the young Weyl to emphasize "sociological" factors. Indeed, when Stone announced that he would write a minority report, Courant acknowledged that having the committee unanimously accept the report would be preferable, but added, "It would not do any good to soft pedal what seems to many of us very important aspects of the situation."[55] When it came to applied mathematics, paying the bill was not sufficient for the success of the field. It might have even hurt it, as it set it apart from pure mathematics and only entrenched the views of some that applied mathematics was less prestigious than pure mathematics.

To be fair, the applied mathematicians on the committee did more than just argue for the importance of applied research. They also—and here again Courant's influence was noticeable—criticized pure mathematicians for being too abstract. While the report celebrated modern axiomatics, it also warned that being pushed to an extreme by pure mathematicians could render mathematics futile. The growing tendency toward specialization and axiomatization, the final report proclaimed, had the "danger of a deterioration of mathematics to the level of manipulating esoterically arbitrary symbols." It added, "This is pure mathematics in its most barren form." As if this were not enough, the report distinguished not between pure and applied mathematics, but instead between what it parsed as "monastic" and "secular" mathematics: "*Monastic mathematics*. . . , in turning down the road of completely detached, self-motivated abstraction, renounces any commitment to the particular world in which we live."[56] The report challenged the idea that research in abstract algebra or topology would have any future applicability. Pure mathematicians, in this view, were pursuing knowledge for its own sake with no consideration as to how or whether it might have any future use.

In defending pure mathematical research, mathematicians turned again to abstraction. The portrayal of pure mathematics in the report, Mac Lane added, was misguided and "ignore[d] the sense in which all mathematicians [were] concerned with basic classical problems."[57] These problems, he added, "deal[t] with basic mathematical substance which [was] based on *fact* and on *physics* in exactly the same way as [were] the applications listed in the report."[58] Mac Lane maintained that even though it might not be immediately apparent how category theory related to mechanics, the connection could be established.[59] Long and detailed, Stone's minority report put forward his own vision of mathematical knowledge. "Though concentration of one's attention on the details might at first seem to indicate the contrary," he wrote, "pure mathematics does in fact continue to revolve about the great central problems of number theory, of geometry, and of analysis, which deal with matters fully as concrete as the abstractions of the atomic or nuclear physicist."[60] Stone's argument was twofold. Like Mac Lane, Stone held that classical and modern mathematics were not separate. As long as classical mathematics deserved to be supported, Stone argued, pure mathematics did as well. Moreover, Stone insisted that the abstractions of mathematicians were not unique, but instead were "as concrete as the abstractions of the atomic or nuclear physicists."[61] His reasoning relied upon the new understanding of mathematization, according to which the concepts of physics were defined analytically rather than phenomenologically, and hence were as abstract as those of mathematicians.

Albert eventually managed to broker a truce among the group by convincing the applied mathematicians to remove some of the more "controversial" paragraphs, which discussed the social factors impeding the growth of applied mathematics, from the final report. In the following decade, as funding for mathematics increased, the initial tension between pure and applied mathematicians subsided. Nonetheless, as the number of graduate students in mathematics concomitantly increased exponentially, pure mathematicians continued to dominate academic mathematics. They did so, at least partially, by doubling down on the utility of abstraction. Modern abstraction, they argued, was the source of the field's growing applicability. The abstract nature of high modernist mathematics did not prevent it from being useful, but rather was wholly responsible for its utility.[62]

On the Unreasonable Effectiveness of Mathematics

In 1959, physicist Eugene Wigner delivered a lecture at New York University titled "The Unreasonable Effectiveness of Mathematics in the Natural Sciences." In his speech, which was published and publicized a year later, Wigner asked what might explain mathematics' incredible ability to describe natural laws. "The enormous usefulness of mathematics in the natural sciences," he said, "is something bordering on the *mysterious* and . . . there is no rational explanation for it."[63] No rational explanation, Wigner insisted, could account for the utility of abstract mathematical concepts in physics. This fact, according to Wigner, was not merely a curiosity. Rather, it was nothing less than "the empirical law of epistemology."[64] To truly appreciate the novelty of Wigner's claim it is enough to circle back to the work of the logical positivists, for whom Wigner's "empirical law of epistemology" would have been heresy.

In 1933, more than a quarter century before Wigner would address his audience at NYU, Hans Hahn, mathematician and early member of the Vienna Circle, approached the same problem from a different perspective. In "Logic, Mathematics, and Knowledge of Nature," Hahn similarly asked what the relationship was between mathematics and natural laws. Like

Wigner, Hahn found the idea that preconceived mathematical concepts could account for natural laws to border on the numinous: "The idea that thinking is an instrument for learning more about the world than has been observed, for acquiring knowledge of something that has absolute validity always and everywhere in the world, an instrument for grasping general laws of all being, seems to us wholly mystical."[65] Yet rather than elevating this proposition into an epistemological rule, as Wigner would, Hahn's goal was to discredit it: "Why should what is compelling to our thought also be compelling to the course of the world? Our only recourse would be to believe in a miraculous pre-established harmony between the course of our thought and the course of the world, an idea which is deeply mystical and ultimately theological."[66] Like his fellow logical positivists, Hahn held that there was no place for metaphysics in a theory of knowledge and, hence, nothing could have been more fundamentally incorrect than to place mystery at the heart of epistemology.

A theoretical physicist, Wigner did not approach the question with the same philosophical rigor as Hahn had. However, the discrepancy in their views was not simply due to their respective idiosyncrasies. Rather, it reflected broader intellectual transformations that had taken place in the ensuing decades. Both Wigner and Hahn were interested in the relation between mathematical ideas and theories of the natural world, yet the core question that occupied each one was inherently different. Whereas Hahn queried the relation between *theory* and *observation*, Wigner wanted to know how "pure" mathematics could find application in physical laws. Both men were concerned with the applicability of mathematics, but their interests reflected different social and intellectual concerns.

Philosophers and various scientists have commented and extended upon Wigner's thesis since its publication in 1960, but what has been less remarked upon is the way in which Wigner's ideas were quickly adopted by mathematicians in the 1960s to defend the growth of pure mathematics. "The miracle of the appropriateness of the language of mathematics for the formulation of the laws of physics," Wigner concluded his essay, "is a wonderful gift which we neither understand nor deserve."[67] Mathematicians could not ask for a better champion than Wigner. In the following years, they began pointing directly to his "miracle" to justify the continued support of the federal government for their studies, regardless of how abstract they were.

The notion that mathematical ideas that were developed independent of nature would nonetheless be useful to the theories of physics predates

Wigner. They had most recently been reinvigorated by both Einstein's general relativity theory and quantum theory. This was, after all, the impetus for Hahn and the logical empiricists' philosophy of science.[68] What Wigner did was update the underlying quandary for the postwar years, offering mathematicians a rationale with which to argue for abstract mathematical research. If the applicability of mathematics was a mystery, then not only were mathematicians justified in pursuing mathematical abstraction, but there was also no real pressure to emphasize applied mathematics. This is not to say that this was Wigner's intention. But at least in some mathematical circles, this was how his thesis was interpreted.

Wigner did not directly comment on the abstract nature of contemporary mathematics, but in 1966, Albert, Mac Lane, and three of their colleagues at the University of Chicago made the connection. The occasion was the publication in *Science* of a speech by Richard Hamming, a mathematician working for Bell Laboratories and a well-respected pioneer in the then-burgeoning field of computing. Entitled "Numerical Analysis vs. Mathematics," the article suggested that practitioners of numerical analysis should align their work with scientists rather than mathematicians.[69] In its objectives and methods, Hamming argued, numerical analysis differed from modern mathematical taste. While the emphasis in numerical analysis was on the *methods* used and their clear articulation, in mathematics it was on the precise and rigorous demonstrations of *results*.

Hamming clearly indicated that his goal was not to criticize mathematicians, but rather to draw attention to the differences between the two activities. Yet this was no comfort. Not long after the article was published, Albert, Mac Lane, and their colleagues shot off a letter to *Science* protesting Hamming's "attack" on mathematics.[70] One paragraph in particular had angered these five Chicago mathematicians. Describing contemporary mathematical research, Hamming noted that, unlike in classical times, the postulates of contemporary mathematical theories no longer bore any clear relation to observational data: "It is difficult to imagine how by appeal to observations many of the postulates of current mathematics could either be verified or shown to be unsuitable, and one can only conclude that much of modern mathematics is not related to science."[71] The Chicago mathematicians were exasperated. Such a claim, they believed, was a grievous and damning misrepresentation of contemporary mathematical research.

To correct this false depiction, the five authors began by outlining the basic idea of modern axiomatics. The measure of an axiomatic system, they explained, was the new information it provided about the context from which it arose, but this context did not need to be, they insisted, a system of "material objects." On a higher level, acts and processes such as counting or measuring could also serve as the underlying context for a mathematical theory. In fact, this process could keep going indefinitely: "Once crystallized in a definite form and proved fruitful, the acts and processes and objective difficulties may provide the context for the creation of a new mathematical theory on a higher level by a new act of mathematical abstraction."[72] In their view, it was these successive acts of abstraction that fueled mathematical research.

By describing mathematics as an ongoing process, the Chicago mathematicians seem to have corroborated rather than disproved Hamming's claim: in its most abstract reaches, the postulates of mathematics were indeed not privy to observational verification. However, they had intended the opposite claim. The nonobservational nature of some mathematical theories did not mean that mathematics was not "related" to science. Far from it, they claimed; it was exactly this process of "abstraction piled upon abstraction" that "ha[d] made mathematics a significant tool and a dynamic force in the development of the physical sciences."[73] The divorce of mathematics from its close bond to reality, in their view, only increased its potential utility in science. They invoked Wigner to support their claim: "It is a paradox which lies at the heart of what Wigner has called 'the unreasonable effectiveness of mathematics in the natural sciences.'"[74] High modernist mathematical rewriting was not an impediment to mathematical utility, but its cause.

Five years earlier, in January 1961, Stone was invited to give a talk at the Annual Meeting of the Association of American Colleges in Denver, Colorado. The association invited two representatives of the social sciences (Ben W. Lewis and David B. Truman), two of the humanities (Brand Blanshard and J. N. Douglas), and two of the natural sciences (Stone and John A. Wheeler) to provide an overview of their respective fields, oriented toward teaching undergraduates. Stone offered a broad overview of modern mathematics and called for an overhaul of the education system in light of the growing influence of science on any aspect of human life. Stone's vision of mathematics was all-encompassing. "Mathematics," he explained, "has to be regarded as the corner-stone of all scientific thinking and hence the intricately articulated technological society we are busily engaged in building."[75] Stone emphasized that contemporary mathematics had been characterized by an emphasis on abstraction defining mathematics as the "study of general abstract systems." While he did not refer to Wigner directly, he, too, remarked that the fact that these abstract systems often served as "models for portions of reality" was "kind of *accidental* and to a certain extent *arbitrary*."[76]

The most exciting development in the twentieth century, Stone continued, had been the increasing penetration of mathematics into not just the natural sciences, but the human sciences as well. As new "situations" came under the purview of mathematical analysis, the true power of mathematics, he claimed, became clear: "It may seem to be a stark paradox that, just when mathematics has been brought close to the ultimate in abstractness its applications have begun to multiply and proliferate in an extraordinary fashion."[77] However, Stone assured his listeners that, "far from being paradoxical," the two opposing trends in the development of mathematics had the same source: "For it is only to the extent that mathematics is freed from the bounds which have attached it in the past to particular aspects of reality that it can become the extremely flexible and powerful instrument we need to break paths into areas now beyond our ken."[78] Here, Stone expressed in the most direct way the theory of knowledge that came to dominate some mathematical circles after the war. Mathematical theory was independent of the physical world, but this did not entail that it was not applicable to theories of the world. The opposite was true—it was the abstract nature of contemporary mathematics that made it so useful.

It would be easy to dismiss Stone, Albert, Mac Lane, and other pure mathematicians as cynically attempting to ensure funding for mathematics, but this would miss how widely the belief in the utility of abstraction had become accepted among mathematicians. The problem mathematicians faced was profound: What accounted for the unprecedented applicability of mathematics? And how could it be squared with the continual turn toward high abstraction by pure mathematicians? In the mid-1960s, the mathematical community resolved to answer these questions. The occasion was once again a national report, this time one which aimed to forward a holistic vision of all of mathematics, not just applied mathematics. The Committee on Support of Research in the Mathematical Sciences was, by all accounts, a massive undertaking. The committee commissioned surveys on a range of topics from undergraduate to graduate education and employment opportunities. In addition to the committee's official members, which included Joachim Weyl and John Tukey, approximately forty-five additional highly respected mathematicians took part in the work.[79] When all was said and done, they produced a long report, two additional surveys on education, and an edited volume of essays on modern mathematics.

In many ways, the final report and the work of the committee testify to the great changes the American mathematical community had undergone since the early 1950s. The report can be read as a conscious attempt to avoid the battles of the 1950s, as is evident in its choice of terminology. The report dismisses altogether the division between pure and applied mathematics or, more accurately, the notion of pure mathematics. Instead, the authors suggest the term "core mathematics" to refer to traditional disciplines such as logic, number theory, algebra, geometry, and analysis: "The name 'pure mathematics' is unfortunate since it implies a monastic aloofness from the world at large and an isolation from its scientific, technological, and social concerns. Such an aloofness may be characteristic of some mathematicians. It is certainly not characteristic of mathematics as a collective intellectual endeavor."[80] "Core" located pure mathematics at the very heart of the scientific project.

Announcing that mathematics now played a part in every domain of modern life, the report asserted that the postwar period had witnessed nothing less than the "mathematization of culture."[81] Faster airplanes, longer-ranging rockets, better-designed gas turbines and radio antennae, and more accurate prediction of satellite orbits were all celebrated as examples of mathematics-based technologies. Examples abounded, and the message the report conveyed was clear: mathematics was foundational to the life of the nation. Despite this utilitarian view of mathematics, when it came to defining what mathematics was, abstraction was the answer: "In a simplified way mathematics consists of abstractions of real situations, abstractions of abstractions of real situations, and so on."[82] It then added, "It is *surprising* but *true* that these abstractions of abstractions often turn out to further our knowledge and control of the world in which we live."[83] I point to this report because it demonstrates that these ideas were not limited to the musings of one or two mathematicians, but also appeared in the culmination of an attempt by a wide cross- section of the mathematical community to arrive at some consensus as to what mathematical knowledge was.

Applied mathematicians similarly adhered to Wigner's mystery. Rees, who advocated on behalf of applied mathematics throughout the postwar period, concluded in an essay on the nature of mathematics that the "realms conquered by mathematics solely because of their intrinsic interest to mathematicians ha[d] provided in the part, and continue[d] to provide, parts of the conceptual framework in which other scientists view[ed] their world."[84] In an essay on "the applicability of mathematics," Stanislaw Ulam, who is most famous for his work on the design of the thermonuclear bomb, also advanced an abstractionist theory of knowledge. He wrote, "Current research in mathematics tends toward ever-more-varied abstraction. Yet the most far-reaching excursions into mathematical theory may lead to applications not only within mathematics itself but also in physics and the natural sciences in general."[85] True, Ulam conceded, much of published mathematics was esoteric and specialized, but one should think of those papers as " 'patrols' sent into the unknown in all directions." He then added, "Philosophically it is curious that mathematical idealizations which at first sight seemed 'irrational' have led to the most useful and practical consequences."[86] That both pure and applied mathematicians gestured toward Wigner's unreasonableness argument in order to explain the applicability of mathematics indicates how pervasive his idea had become by the late 1960s.

However, in appropriating Wigner's ideas, mathematicians also amended his argument and expanded it beyond its original context. The so-called mystery of mathematics was no longer confined to its function as a descriptive language, but now extended to both prescriptive theories and technological innovations. This revision of Wigner's ideas was apparent in the 1968 report which, among the "surprising" applications of pure mathematics, included not only the customary examples of the use of differential geometry in relativity theory and matrix algebra in quantum mechanics, but also the use of Boolean algebra in computer architecture. Moreover, it seems that, to the authors of the report, what was surprising was not that independent mathematical concepts found expression in objective natural laws, but the ubiquity of the phenomena, which they now read into every application from engineering to communication technologies and aeronautics.

In 1969, Joachim Weyl coedited *The Spirit and the Uses of the Mathematical Sciences* with operations researcher Thomas L. Saaty. Opening their introduction, they wrote, "Everyone, layman and professional alike, who has been in touch with mathematics in recent years had been deeply impressed by the extent to which its forms of reasoning have spread into practically all aspects of modern life."[87] This growth, they added, was the outcome of two forces—on the one hand, development internal to the mathematical sciences, and on the other, its "*inevitable* application (its 'unreasonable effectiveness,' as Eugene Wigner's article says) in science and technology to produce dams, power grids, innovations in medicine, new crop varieties, and a multiplicity of other items required by the world today."[88] Indeed, the two editors chose to reprint Wigner's article in their book alongside articles on the uses of mathematics in risk assessment, computers, operations research, and economics.

The difference might not seem to have amounted to much at first, but it was significant. Claiming that it was puzzling that autonomous mathematical ideas manifested in laws of nature that presumably existed independent of human thought was one thing; arguing that it was remarkable that independent mathematical ideas appeared in contemporary technologies was another matter. After all, computers, communication technologies, and ballistic rockets do not exist independent of human thought. They are by definition the creations of human minds. A deterministic perspective—that is, the belief that all of the contemporary developments described in the report were inevitably as they are—is the unstated assumption underlying the mystery thesis. As if it were absolutely necessary that modern computer architecture would depend on Boolean algebra! Wigner knew this. Having stated his epistemological law, Wigner was quick to ask whether the surprising appearance of independent mathematical concepts in natural laws meant that they could be stated otherwise: "It is just this uncanny usefulness of mathematical concepts that raises the question of the uniqueness of our physical theories."[89] If our natural laws could be expressed mathematically in more than one way, the "mystery" would not be so enchanting. It is only if you believe the laws of nature exist independent of human thought that their mathematical expression demands awe, but can the same be said of a power grid?

Historians of science have long agreed that the division between basic and applied research is constructed and that any attempt to draw clear distinctions between the two is bound to fail. What postwar mathematicians did was practically dismantle the notion of utility in mathematical research. The case for pure mathematics had to be made. During World War II the scientific elite of the country were not enamored with the leaders of the mathematical community. To ensure pure mathematics would be included among the new fields vying for military and federal support, these leaders had to stretch and question the notion of utility itself, and they succeeded. At a time when scientific mobilization in the country pointed in one direction, pure mathematicians stubbornly refused to fall in line with the majority. Not only were they able to remain in the ivory tower, but their success also impeded the growth of applied mathematics. This was the true paradox.

The Education of the Applied Mathematician

The irony is that Wigner delivered his famous lecture as the inaugural Richard Courant Lecture at NYU. The lecture series had been founded a year earlier by some of Courant's friends and colleagues in celebration of his seventieth birthday. The ideas expressed in Wigner's lecture would not only help pure mathematicians maintain autonomy over their field, but also stifle the growth of applied mathematics. The belief that the real strength of mathematics was its power of abstraction remained a hurdle applied mathematicians had yet to surmount. No one was more cognizant of this fact than Courant. After he read a published version of Stone's lecture to the Association of American Colleges, he took Stone to task during a meeting of applied mathematicians. Stone's article, Courant explained, attributed the success of modern mathematics to its detachment from reality and its abstract nature. Lightly caricaturing Stone's argument, Courant announced, "Thus, the mathematical mind, freed from ballast, may soar to heights from which reality on the ground can be perfectly observed and mastered."[90] Courant did not disagree with Stone on that point. Rather, he believed that, especially as they applied to education, Stone's ideas were simply one-sided. "The danger of enthusiastic abstractionism," he wrote, "is compounded by the fact that this fashion does not at all advocate nonsense, but merely promotes a half-truth. One-sided half-truths must not be allowed to sweep aside the vital aspects of the balanced whole truth."[91] The abstractionist theory presented by Stone was not incorrect, but neither was it the whole story. What was necessary, on Courant's account, was to counter this abstractionism with a mathematics rooted squarely in reality: "Abstraction and generalization [are] not more vital for mathematics than individuality of phenomena and, above all, not more than intuition."[92] Courant wished to focus on the interplay between the two.[93]

The problem was that besides the NYU group, there were not many institutions in the US where the abstract approach was balanced by an emphasis on mathematical theory rooted in physical reality. Even during the 1960s, when the overall growth of mathematics accelerated, applied mathematics remained marginal. For example, a 1961 survey found that out of ninety-five American institutions offering graduate training in mathematics, only forty-nine offered specialization in applied mathematics.[94] In comparison, eighty-six institutions offered specialization in analysis, seventy-seven in algebra, and sixty-seven in topology. Four years later, there were only four departments across the country dedicated specifically to training applied mathematicians. At the same time, computer science, statistics, and operations research matured into autonomous disciplines, with their own set of standards and professional associations. The first computer science department was founded in 1962 at Purdue University, and by 1967 there were twenty-four such departments across the country.[95] Applied mathematics, as a coherent research field, failed to take hold within academic departments. Pure mathematicians were happy to point to the numerous applications of mathematics, but they were not eager to develop them themselves.

The problem with Wigner's thesis was not just that it provided justification for pure mathematicians' research, but that it also called into question what training in applied mathematics should look like. This issue came to the fore during a conference dedicated to education in applied mathematics in 1966. The conference included many leaders in the field, among whom were C. C. Lin (MIT), Tukey (Princeton), William Prager (Brown), Courant (NYU), and George Dantzig (Stanford). As in the previous decade, it was difficult to arrive at a clear consensus as to what applied mathematics entailed. Over the course of the conference, attendees debated whether the applied mathematician represented a distinct professional identity and if so, how it differed from the pure mathematician, the theoretical physicist, and the engineer.

During the second day of the conference, Wigner's predicament sparked discussion. Werner Rheinboldt, an applied mathematician from the University of Maryland, gave a commentary on the topic of the mathematical curriculum. Rheinboldt focused his discussion on three areas of research that in recent years had become increasingly pertinent in computing: automata theory, theory of formal languages, and graph theory and combinatorics. Having briefly described each of these fields, Rheinboldt concluded by noting that "not one of the three fields discussed here was originally thought of as having a direct connection with applied mathematics, and their development before the advent of computers took scarcely any notice of application." Moreover, he added, "Recent 'applied' developments in these fields still are often regarded as being outside their mainstream activity."[96] When general discussion began soon thereafter, the first to speak was Peter Lax. A Hungarian émigré, Lax had earned his PhD from NYU and remained teaching at the institution throughout his career.

Lax took issue with Rheinboldt's characterization of the relation between pure and applied mathematics: "I disagree with you in this business of how applications interact with mathematics. Maybe I am misinterpreting your remarks; they seemed to say this is best left to randomness. But as C. C. Lin remarked, there is a certain filter which influences the probabilities. I emphatically do not think it should be left to the random subconscious of the mathematical mind, but it is something that has to be fostered most deliberately."[97] Lax had perfectly encapsulated what was at stake. If the relation between mathematical theories and the natural world were, as Rheinboldt suggested, unpredictable, then nothing could be done to train applied mathematicians. After all, it was the job of the applied mathematician to draw connections between abstract mathematics and the physical world. And the goal of training was to teach students how to do exactly that.

This was a real point of contention. Lipman Bers, who had been Lax's colleague at NYU but had recently decamped uptown to Columbia University, replied to Lax, "I would like to first emphasize a point [on] which we seem to disagree. This is this matter of surprises, the cases of unexpected influence of mathematical discoveries in other fields." He then added, "I do not believe that we should try to program education for surprises. A surprise is, by definition, something which cannot be programmed for."[98] Bers did not believe that there should not be a training program in applied mathematics, but since it was impossible to predict which mathematics would be useful in the future, the goal of training in the field would be to provide students with necessary flexibility.[99]

For Bers, the mystery of mathematics was not a matter of belief, but rather a matter of historical fact. "It is trite," he added, "to talk again about what happened to Einstein when he needed tensor calculus, but it is a fact, it *did* happen." Bers noted that recognizing that this was what happened in the past does not tell us why it happened, but this was beside the point. He further suggested that the same "spirit of unity of mathematics should permeate . . . the education of applied mathematicians. This means that the future applied mathematicians . . . should be trained in the kind of abstract thinking which helps unify mathematics."[100] The strength of mathematics, pure or applied, was, according to Bers, abstraction.

Even when they considered the most mundane aspect of education—which courses to teach and how to integrate research and education—mathematicians found it impossible to avoid a more philosophical discussion of the theory of knowledge production in mathematics. This fact was not lost on the participants of the conference. When George Pólya took to the stage the next day, he confessed that when the presenters on his panel had met the night before, their discussion had turned to philosophy.

"Please, I am not against philosophy, I believe in philosophy," Polya added. However, at some point it was crucial to "get down to brass tacks. We decided yesterday that we shall try to get down to brass tacks, to talk about concrete things. We shall try. Whether we shall succeed is another question."[101] The conversation that followed focused for the most part on concrete programs, but it was impossible not to slip every so often into a consideration of what applied mathematics was.

Lax was not the only one who implored his colleagues to take an active approach to application. Barkley Rosser, who directed the Army Mathematics Research Center at the University of Wisconsin–Madison, returned to the subject a day later in his address on research programs in applied mathematics. Although he was trained as a logician, Rosser had spent the war years conducting research on ballistics. Now, Rosser gave an impassioned speech on the active pursuit of applied mathematics. "You do not sit in your office with a sign on your door saying 'Problems solved here' and sit back and wait for people to come in with problems," he told members of the audience. "We have somehow got to get out in the trenches and run into some good problems."[102] To bolster his argument, Rosser gave as an example mathematician Mark Kac, who, he said, used to walk from one Cornell department to the next asking researchers what they were doing. "He would interest himself in their problems . . . they would tell him, he would get all excited about this; pretty soon he was solving problems, too."[103] Even if there was room for mystery in applied mathematics, mathematicians had to at least know what other scientists were researching.

Rosser touched on one of the main issues that distinguished discussion about mathematical applicability in the 1930s from similar conversations being held in the 1960s. When mathematicians in the 1930s noted that pure mathematical investigation could lead to physical discoveries, they did not have in mind the postwar mathematicians' autonomous conception of the field. In the 1930s, the bond between pure mathematicians and theoretical physicists was much closer than it would be in the three decades that followed. Even though Veblen's and Birkhoff's main contributions were in pure mathematics, they were well familiar with contemporary theories in physics and dedicated much of their energies to the study of general relativity. By the 1960s, the gap between pure mathematics and physics felt unbridgeable. Ulam, who attended the conference, commented on the matter: "I have a feeling if you ask a random mathematician to describe in two sentences, in his own words, the difference between the neutron and electron, in the vast majority of cases you would get a complete blank, hardly any answer, or any interest, as a matter of fact, at all."[104] That Ulam recognized problems of education begins to show how difficult mathematicians found it to simultaneously hold on to their theory of knowledge and its practical implications.

Increased specialization on both sides undoubtedly calcified the split between pure mathematicians and physicists, but mathematicians' abstractionism dominated this bifurcation. Classical mathematics' origin in the lived world was more obvious. Especially for the younger generation of mathematicians trained in the postwar period and steeped from the get-go in abstractionist mathematics, the attachment to classical theory had been completely severed. Their approach to mathematics was abstractionist through and through.

Kac, whom Rosser praised in his speech, was one of the few mathematicians who recognized this predicament and tried to amend it in his own teaching. He did not dispute the fact that abstract mathematical ideas could often prove to be useful in unexpected domains, but he insisted that too often the abstract presentation of a given theory obscured its potential applicability. In 1959, Kac published a book on statistics and probability that reads as both an introduction to the subject and a case against abstractionist mathematics. "In the pages that follow," Kac writes in the preface, "I have tried to rescue statistical independence from the fate of abstract oblivion by showing how in its simplest form it arises in various contexts cutting across different mathematical disciplines."[105] The modern trend toward generality and abstraction, Kac argues, had the effect of obscuring not only the underlying ideas and motivation for the development of a theory but also any clue as to how it might be used.

In one example after another, Kac emphasizes what he terms the "price of abstraction." "Unrestricted abstraction," he writes, "tends also to divert attention from whole areas of application whose very discovery depends on features that the abstract point of view rules out as being accidental."[106] In other words, the promise of the abstract point of view—its ability to home in on the "essential" while ignoring particularities—was also its failing. Having worked in great detail on a problem about the probability that a given mathematical series with random plus and minus signs converges, Kac notes that a certain analogy between two different parts of mathematics emerges. "What is the moral of all of this?" he then asks the reader. "Could this be achieved if we had insisted on treating 'heads or tails' abstractly?" The choice of words here is not accidental. For Kac and other applied mathematicians, the abstractionist theory that dominated pure mathematics endangered the future of the field. It threatened to sever mathematics from the lived world, not only in principle but also in practice. How good could a mathematics theory be if there was no one around to put it to use?

During the conference on the training of applied mathematicians during the postwar period, Bers offered those in attendance another path. Instead of arguing whether or not it was possible to train for "surprises," he suggested mathematicians should look to history: "Let us stimulate research in the history of modern mathematics . . . let us try to find 'what did really happen.' I think if we do we will find that the traditional picture of problems coming from the outside into mathematics, being solved there and then going back, is exceedingly oversimplified . . . The true history of the interplay between applications and pure mathematics is highly interesting and should be studied and taught."[107] Bers believed that the history of mathematics would offer an answer to the current predicament mathematicians were facing. Another former colleague of Bers at NYU, Morris Kline, did exactly that. Kline turned to history in order to repudiate what he conceived as the misguided direction of pure mathematics.

Kline, who published numerous books on the history of mathematics, held that pure mathematicians' elevation of mathematics for its own sake was a misreading of the historical record.[108] According to Kline, once turn-of-the-century mathematicians acknowledged that mathematics was not bound by reality, they were surprised to find that these seemingly arbitrary theories were nonetheless useful in the study of nature (in other words, Kline did not so much answer Wigner's question as invalidate it). While the reason for the utility of these mathematical theories was that they were historically rooted in the study of nature, mathematicians took their successful applicability as a sign that they were "free to create arbitrary structures." Once they recognized that their new theories were "so far removed from even potential application," they began defending them as having inherent value for their own sake.[109] The debate between pure and applied mathematics in the present day, Kline concluded, was rooted in the misunderstanding that arose at the turn of the century.

Kline was not the only mathematician who turned to the history of mathematics to find answers. However, as I show in chapter 6, by the 1970s these mathematicians were not the only ones interested in the history of mathematics. Young professional historians of mathematics were also coming into the field, and the two groups had differing ideas about the nature of mathematics and its history.

CHAPTER SIX

Historical Abstraction

Kuhn, Skinner, and the Problem of the Weekday Platonist

However, you should not expect me to describe the mathematical way of thinking much more clearly than one can describe, say, the democratic way of life.
—Hermann Weyl

The turn of the century transformation in mathematics and the philosophical debates it inspired were as much about the meaning and definitions of concepts as they were about the foundations of mathematics. Turn-of-the-century thinkers across a number of disciplines called into question the abstractionist understanding of concept formation, namely the idea that concepts are formed by a process of identifying common traits across objects and extracting them away while ignoring any particularities. Philosophers and logicians might have first focused on mathematical concepts such as points and lines, but their accounts of how conceptual innovation occurred had implications outside of mathematics. This is most obvious in the work of Gottlob Frege. Frege's first book, *Begriffsschrift* (Concept Notation), published in 1879, represents his attempts to construct a new two-dimensional notional system for predicate logic, which he believed represented "pure thought." Already in his *Habilitationsschrift*, Frege considered the problem of concept extension. He asked how to extend the concept of "quantity," such as length, in light of imaginary numbers. "According to the old conception," Frege wrote, "length appears as something material which fills the straight line between its end points."

Over time and with new developments, however, "the concept ha[d] . . . gradually freed itself from intuition and made itself independent."[1] In *The Foundations of Arithmetic*, Frege then turned his interest to the concept of number.[2]

Frege's investigations into the logical foundation of mathematics also engendered the philosophy of language. As Michael Dummett explains, in his effort to produce a framework within which mathematical proofs could be rigorously established, Frege also analyzed the structure of the statements which made up a given proof. As a consequence, Frege's attention turned not only to the syntactic but also to the semantic analysis of language. As Dummett writes, "Frege has, in other words, to provide the foundation of a theory of meaning."[3] In the following decades, Frege's analysis would be extended and expanded upon by Bertrand Russell and, later, Ludwig Wittgenstein, thereby founding analytic philosophy. Mathematicians, as this book has described, mostly followed a different path—axiomatics.

While mathematicians' embrace of modern axiomatics promised to solve the crisis of referentiality, which had dominated artistic and intellectual activity at the turn of the century, it triggered a different crisis in *meaning*. "Modernist purism," writes Herbert Mehrtens, "restricted mathematics to the construction of strictly regulated worlds of meaning made from typographical sign systems."[4] Up until the 1930s, mathematicians, philosophers, and logicians had continued to debate the source of meaning in mathematics, but by the postwar period most research mathematicians were happy to leave philosophical quandaries behind. Nonetheless, mathematics for them was more than just "typographical sign systems."[5] To search for meaning, and an answer to the question "What is mathematics?" they now turned to history. It was history that could reveal, so they held, the true meaning of abstract mathematical ideas.

Mathematicians' turn to history was not antithetical to high modernist mathematics, but rather a response to it. It was precisely the *force* of contemporary rewriting that made history so potent. It was mathematicians' research commitment to synchronic analysis, which "erased" history in favor of formal axiomatics, that accounted for the fervor with which mathematicians pursued a diachronic approach to their history, emphasizing the evolution rather than the novelty of a given mathematical idea.[6] A mathematical concept might be defined axiomatically, but its meaning and the idea it represented could only be recovered from the historical record. How else might one explain that even Bourbaki, arch-modernist of mathematics, included in its books chapters of historical surveys ranging from Babylonian mathematics to the present?

By the 1960s, however, mathematicians' approach began to be challenged, first by the publication in 1962 of Thomas Kuhn's *The Structure of Scientific Revolutions* and seven years later by Quentin Skinner's publication of his now famous essay "Meaning and Understanding in the History of Ideas."[7] Together, both works challenged mathematicians' project of probing the history of the field in order to answer the question "What is mathematics?" And, notably, both works were influenced by the philosophy of language as developed by analytic philosophers. Joel Isaac points out that Kuhn read widely in linguistics before writing *Structure*. One of the main themes that came from his reading, according to Isaac, was "his criticism of the idea that the relationship between theory and experiment, or between a concept and its application, could be accounted for in the terms of logical positivism or operationalism."[8] What Kuhn would critique was the idea that a pure language of science, one which remained the same across different theories, existed. Logical positivists, who believed that all knowledge must be rooted in experience, held that such pure language, or "protocol statements," would describe immediate experience or sense perception. These protocol statements could then be combined using logic into more complex statements. On the other hand, Kuhn and other antipositivists argued that such protocol language could not exist in principle. There simply were no linguistic terms that retained their meaning outside of a linguistic (or theoretical) paradigm. As Peter Galison puts it, "This was the enemy: a neutral, unproblematic Archimedean point outside of a theoretical structure."[9] Kuhn's philosophical position emerged from and was translated into a study of history. The two were deeply connected.

For Kuhn, each paradigm was characterized by coherent theoretical language, but any attempt to try and translate across paradigms was futile. In 1970, Kuhn explained that, while philosophers since the seventeenth century had assumed that a neutral language recording pure sensations existed, they had "now abandoned hope of achieving any such ideal, but many of them continue[d] to assume that theories [could] be compared by recourse to a basic vocabulary consisting entirely of words which [were] attached to nature in ways that [were] unproblematic and, to the extent necessary, independent of theory."[10] For Kuhn, words were attached to nature only within a given paradigm. For the historian, this meant that scientific concepts could not be assumed to remain the same across paradigms.

Skinner, like Kuhn, was well read in the philosophy of language. In particular, Skinner was influenced by the later Wittgenstein. In *Philosophical Investigations*, Wittgenstein challenged Frege's theory of concepts, arguing that a concept's meaning could not be isolated outside of its use in language. As he famously put it, "The meaning of a word *is* its use in the language."[11] The lesson Skinner took from Wittgenstein was that there was no point in the historian asking what the "true" meaning of a word was. Instead, all the historian could and should do was focus on the way words were put into use in specific texts.[12] Historical contextualism, as practiced by Skinner (and his Cambridge colleagues), can thus be said to be an application of Wittgensteinian philosophy to the practice of history. Skinner, however, did more than just forward a new historical methodology. He also forcefully attacked contemporary historical practice for being rooted in a *philosophical* fallacy. According to Skinner, historians of ideas (consciously or not) held onto the belief that concepts had an essential meaning and could be unambiguously defined. This in turn led them to subscribe to a host of what Skinner termed historical mythologies and develop a tendency to project the present onto the past. Like Kuhn, Skinner's philosophical position and his theory of history were inextricably bound together.

In other words, the two major historiographic reorientations of the 1960s and the 1970s, which, taken in sum, practically put an end to the history of ideas as it had been practiced in the 1940s and 1950s, were rooted in a critique of the semantic tradition that started with an inquiry into the *foundation of mathematics*, beginning with Gottlob Frege and proceeding with Bertrand Russell's efforts to secure the logical ground of meaning. If Kuhn convinced historians of science that it was impractical to try to evaluate a scientific concept before and after a scientific revolution, Skinner extended this critique to include ideas tout court. Neither energy nor liberty had any essential meaning that the historian could attempt to excavate by following the historical record. Yet the search for the "true" meaning of an idea was exactly what mathematicians were after in their historical studies.

In what follows, I trace mathematicians' turn to history and the ensuing debate between historically minded mathematicians and historians of mathematics in the 1960s and 1970s. It would be easy to dismiss the mathematical community as either insular or simply ahistorical (although surely they were guilty at times of both). In this chapter, however, I want to resist such temptations and ask instead what can be gained by following the 1970s debates while taking seriously mathematicians' historical studies. Mathematics here is key. While starting in the 1970s, one of the greatest sins an intellectual historian could commit was holding on to a Platonist understanding of ideas; among mathematicians (and some philosophers) Platonism was in fact the norm.[13] Of course, this is not to say that mathematical ideas are timeless. There is abundant literature to the contrary. Rather, the 1970s historiographic debates among historians of mathematics and mathematicians reflect a deeper historiographic concern common to all intellectual historians about the complicated relationship between ideas, concepts, and temporality.

Mathematicians' histories might be anachronistic, but they exemplify a tension common in *any* history of ideas—that between the *permanent* and the *transitional*. The problem is one that intellectual historians have been grappling with since the 1970s, and it is especially evident in more recent calls for a return to a history of ideas. This is why conceptual historians and historians of ideas find themselves debating what are fundamentally ontological questions, and why a history of concepts necessitates a history of temporality. I point to the history of mathematics to ask whether a history of ideas that recognizes continuity over time is possible even without any conscious or unconscious belief in an abstract realm of timeless ideas.

I begin by outlining how mathematicians' growing interest in the history of mathematics in the postwar moment collided with that of historians of mathematics starting in the 1970s. While the criticisms historians of mathematics have directed toward mathematicians were in line with those voiced by historians of ideas at the time, the dispute, I argue, had less to do with the nature of history than with the nature of mathematical practice. Taking mathematicians and historians as interested in two related yet distinct projects, I suggest that mathematical history understood as a *mathematical* activity offers one way to hold on to historical continuity without essentialism, for it was continuity in practice rather than ontological continuity that characterized mathematicians' work.

Between the History of Science and Mathematics

In 1974, mathematicians and historians of mathematics convened at a two-day workshop on the history of modern mathematics hosted under the auspices of the American Academy of Arts and Sciences. The workshop was organized by a committee that included mathematicians Garrett

Birkhoff (chair), Felix E. Browder, Kenneth May, and historians I. B. Cohen and Thomas Hawkins. The stated goal of the conference was to draw the attention of both mathematicians and historians to the development of mathematics after 1800, a period which the organizers believed was underresearched. Recognizing the task before them, the organizers made a point to include both established mathematicians and historians and younger colleagues who were just entering the field. One of the only two women speakers in the conference, Elaine Koppelman, had received her PhD in the history of science from Johns Hopkins University in 1969. She had first pursued a doctoral degree in mathematics at Yale University, but decided to change her studies and as such represented the growing professionalization of the history of mathematics as an independent field. At the conference, Koppelman presented a potential scheme for progress in mathematics (see Fig. 6.1). She suggested, for example, that some new mathematical results were arrived at by transferring a technique from one context to a new one, or by combining two existing fields into a new one. This classification, Koppelman suggested, could help shed light on particular historical events. It offered a framework from which to evaluate the development of mathematical ideas.

After she concluded her talk, philosopher Hilary Putnam interjected. What Koppelman was offering was taxonomy without theory, he objected. Putnam noted that while he had many disagreements with Kuhn's work, at least Kuhn used his categorization of periods of paradigm shift versus normal science to offer a theory of the development of science. Was Koppelman's goal simply to offer a taxonomy, or to propose a more general law of the growth of mathematics? Koppelman defended her paper, explaining that the goal was to offer a framework with which specific cases could be analyzed. However, Putnam remained skeptical and took the opportunity to voice his general dissatisfaction with the historians in the group: "The historians of mathematics here today have talked entirely in generalities, and the mathematicians here are *doing* the history of mathematics. I see two possible explanations: (1) defensiveness, a mistake which spoils the possibility of interchange here, and (2) that the example of Kuhn has spread so, that now every historian has to produce generalities."[14] Putnam's interjections begin to point to the problem that historians of mathematics were facing in the 1970s. On the one hand, they had to contend with mathematicians, who in the postwar moment increasingly turned to history to make sense of their field. On the other hand, they had to reassess the relation between the history of mathematics and the

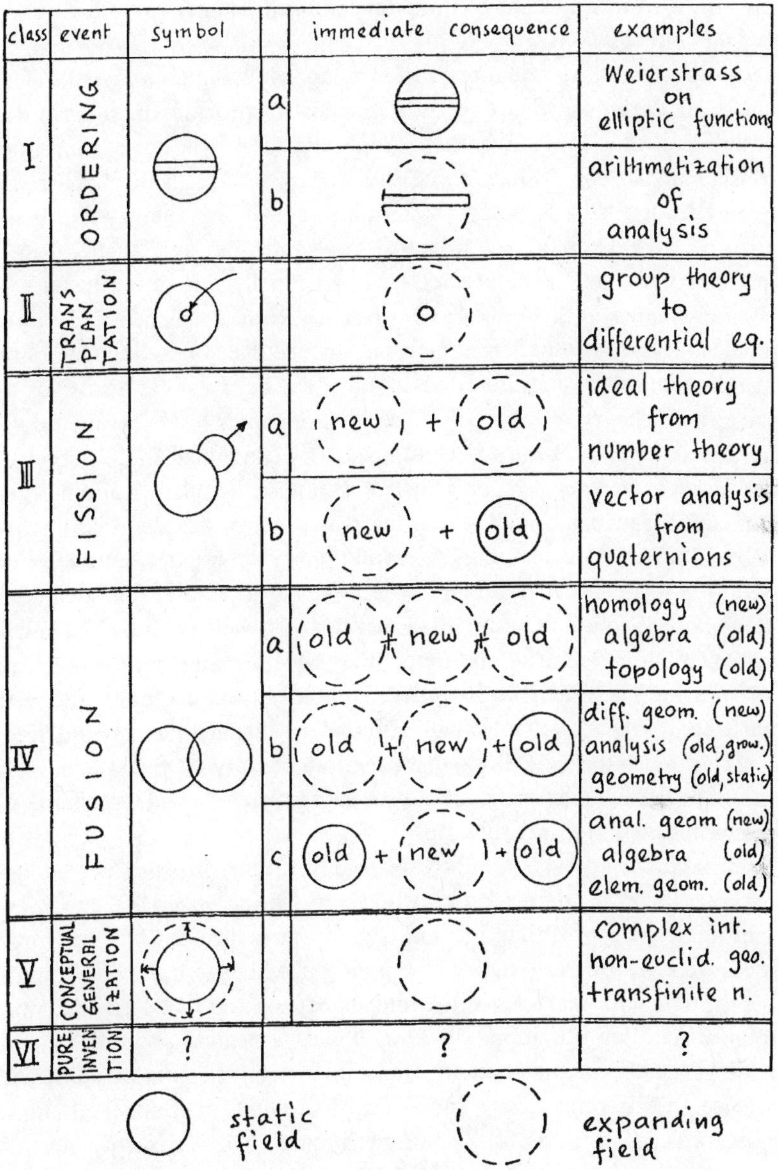

class	event	symbol		immediate consequence	examples
I	ORDERING		a		Weierstrass on elliptic functions
			b		arithmetization of analysis
II	TRANSPLANTATION			o	group theory to differential eq.
III	FISSION		a	new + old	ideal theory from number theory
			b	new + old	vector analysis from quaternions
IV	FUSION		a	old + new + old	homology (new) algebra (old) topology (old)
			b	old + new + old	diff. geom. (new) analysis (old, grow.) geometry (old, static)
			c	old + new + old	anal. geom (new) algebra (old) elem. geom. (old)
V	CONCEPTUAL GENERALIZATION				complex int. non-euclid. geo. transfinite n.
VI	PURE INVENTION	?		?	?

static field

expanding field

FIGURE 6.1. Koppelman's classification of mathematical progress.

Source: *Historia Mathematica* 2, no. 4, 1975

history of science in light of the publication of Kuhn's *The Structure of Scientific Revolutions*.

Some of those in attendance tried to engage with Koppelman's classification, asking to what degree it applied to various mathematical developments, but the conversation kept returning to Putnam's critique. Mathematician-turned-historian Kenneth May noted, "This emphasis on generalization is quite a new thing in history," and mathematician Felix Browder added that one "could transform Koppelman's classification into a Kuhnian system" if one wished to do so. Finally, one of the historians of mathematics in the audience, Michael J. Crowe, remarked that it was ironic that the philosopher in the room was the one counseling historians against theoretical considerations. "There is a need among historians to have some theory in terms of which to order their historical material," he added.[15] He too returned to Kuhn. Crowe explained that for him to designate some event or period as revolutionary entailed "not only that something new came in, but also that some entity was deposed or discarded."[16] With this meaning of "revolutionary" in mind, would you say, he asked Koppelman, that there have been revolutions in mathematics? Koppelman replied that she disagreed with Crowe's definition of revolution. While it was true, she noted, that once a mathematical theorem was proven, it could not be discarded, a mathematical revolution occurred when the "scope" and "meaning" of old ideas changed. "The world 'line' does not mean the same thing to a twentieth-century mathematician as it meant to Euclid," she argued, "and the acceptance of a wider meaning of the term constituted a revolution."[17]

It is difficult to overstate the influence of Kuhn's *Structure* on the history of science. Alexander Koyré, who Kuhn acknowledged had an important influence on his thinking, had already demonstrated how scientific ideas were wedded to religious and philosophical ones, but it was Kuhn's description of normal science, the role of pedagogy, and the scientific community that drew attention to the *sociological* aspects of scientific knowledge. In the aftermath of Kuhn's publication, social and political studies of science grew in prominence, and Whiggish accounts of science fell out of fashion. The transition was not necessarily smooth. In the history of medicine, for example, tension arose between the younger generation, who wished to integrate lessons from social history, and the older generation, who felt that they were writing "medical history without medicine."[18] The history of mathematics was similarly embroiled in debates between internalists and externalists.[19] While the 1974 conference attempted (unsuccessfully) to bridge the chasm between historically minded mathematicians and historians of mathematics, the period was one of heated debates between the two groups. During the conference, for example, Bourbaki founder Jean Dieudonné questioned the relationship between mathematics and factors external to its development, stating, "The general history of the seventeenth century has no connection with Fermat's theory of numbers."[20] His assertion was quickly challenged by some of the more historically oriented participants.

The problem that historians of mathematics were facing, however, was much greater than that. What Kuhn's theory drew attention to was the incompatibility between the *historicity* of science and that of mathematics. Kuhn explained the implications of his theory for scientific historicity when he tried to account for the difference between science and art. Scientists and artists, Kuhn claimed, had "sharply divergent responses to their discipline's past." Contemporary artists might have different "sensibilities" than their predecessors, but "the past products of artistic activity [were] still vital parts of the artistic scene." As he explained, the works of Picasso did not in themselves relegate Rembrandt's paintings to dusty storage closets. Not so in science: "In science new breakthroughs do initiate the removal of suddenly outdated books and journals from their active position in a science library to the desuetude of a general depository."[21] Or as he forcefully put it, "Unlike art, *science destroys its past.*"[22] Here again, mathematicians aligned themselves more closely with artists than with scientists.

Not only were mathematical truths, once established, neither refutable nor discardable, but more fundamentally many mathematicians understood their work to be a continuous dialogue with past mathematicians. Throughout the postwar period, mathematicians compared the mathematical library to the scientific laboratory, arguing that a well-stocked library was instrumental to mathematical research. Dieudonné captured this sentiment perfectly in the discussion proceeding Koppelman's presentation, when he explained, "A mathematical theorem can never be thrown out if it has been proved true. But it may become insignificant. On the other hand, it may look insignificant and in later years become a most fundamental thing."[23] At least in principle, the mathematical past could always speak to the present.

When it was Crowe's turn to present, he returned to the applicability of Kuhn's theory to mathematics. Crowe offered ten "laws" that accounted for conceptual change in mathematics. The tenth law simply stated:

"Revolutions never occur in mathematics."[24] Crowe, who had received his doctorate in the history of science in 1965 from the University of Wisconsin, acknowledged that Koppelman's paper and his shared similar goals, "attempting to discern patterns of change in the history of mathematics." However, he amended that they differed when it came to the question of revolution: "My denial of their existence is based on a somewhat restricted definition of 'revolution,' which in my view entails the specification that a previously accepted entity *within* mathematics proper be rejected."[25] The discovery of non-Euclidean geometry was not, according to Crowe, a revolution, because its acceptance did not entail a rejection of Euclidean geometry. To be sure, non-Euclidean geometry led to "a revolutionary change in views as to the nature *of* mathematics . . . but not *within* mathematics itself."[26] Crowe and Koppelman's disagreement could be glossed as two historians with differing opinions on the nature of revolutions, divided as to whether mathematics had to destroy its past in order to have a revolution, but such a reading would miss the point. The source of their disagreement came from the distinction Crowe drew between ideas *within* mathematics and those outside of it. Although where and how the line between the two was drawn remained unclear.

After the proceedings of the conference were published in *Historia Mathematica*, German historian of mathematics Herbert Mehrtens took Crowe to task in the journal's pages. Mehrtens reflected on the applicability of Kuhn's analysis to mathematics, asking, Does normal mathematics exist? Are there exemplars and anomalies in mathematics? And what about mathematical paradigms? However, one of the most pertinent aspects of his analysis was a critique of Crowe's use of the preposition "in." To illustrate his point, Mehrtens described how nineteenth-century British mathematicians switched from a Newtonian to a Leibnizian notation of the calculus. As he noted, Cambridge and Oxford scholars strongly resisted the adoption of Leibniz's differential notation, and when it finally happened, the change reoriented British mathematical research. By all accounts, Mehrtens argued, this constituted a revolution for the British mathematical community. Crowe might argue otherwise, he added, only by insisting on the preposition "in." "Unfortunately," he wrote, "he does not explain what *in* mathematics means, except that nomenclature, symbolism, metamathematics, methodology, and historiography are not *in* mathematics. Probably Crowe had the 'contents' or the 'substance' of mathematics in mind (what is this?)"[27] What Mehrtens pointed out was that the answer to the question of whether there were revolutions in mathematics depended on one's philosophical position. While Mehrtens did not say so explicitly, what he made clear was that distinguishing between the "content" of mathematics and the symbolism, methodology, and metamathematics was only possible if the historian subscribed to some Platonist notion about the existence of mathematical objects or at least believed that a mathematical idea existed independently of how it was presented—and independently of the community of mathematicians who studied it.[28]

Not only did Kuhn's theory of temporal change in science not accord with that of mathematicians (and some historians of mathematics), but, moreover, many mathematicians outright objected to such generalizations. After another historian of mathematics, Thomas Hawkins, discussed the nature of progress in mathematics during the 1974 conference, mathematician Morris Kline felt the need to defend the historians in the audience: "They came today, not to tell us the history of mathematics, but to discuss the problems of history."[29] Commenting that the mathematicians' talks were very detailed, he added, "The purpose of this meeting was not so much to get the history down from the mathematicians, who would be the best source, but rather to discuss more the problems of how we can all co-operate."[30] Similarly remarking on the difference between the mathematicians' and historians' presentations, Garrett Birkhoff suggested that mathematicians were so attentive to details and close studies because "historical abstractions tend[ed] to go over their heads."[31] He added that discussions between mathematicians and historians of mathematics should be limited to specific examples, with a minimum of philosophical superstructure."[32] In the decade that followed, the history of mathematics continued to be squeezed between mathematics and the history of science, never fully integrating into either.

That Kuhn's book contributed to this disciplinary split is ironic. Kuhn shared logical positivists' desire to explain how science worked and what made it the activity it was. Defending his historical approach in 1970, Kuhn wrote, "I am no less concerned with rational reconstruction, with the discovery of essentials than are philosophers of science."[33] Kuhn believed that there was an underlying *structure* to scientific activity, namely the cycles between normal science and revolutions. There was an "essential" truth about scientific activity that was hidden beneath the surface. The difference was that while logical positivists turned to philosophy to uncover this truth, Kuhn sought it in the historical record. Indeed, in this respect, his theory bears the mark of its time.[34] Peter Gordon suggests that this might explain why Kuhn's work "captured the imagination" in ways other theories did not. "Part of the explanation," Gordon writes, "may be that it is understanding of the underlying 'structure' of historical change—as sudden rather than gradual—confirmed the basic lessons of the other intellectual revolution that was sweeping across the social sciences in Europe: structuralism."[35]

Eight years after Lévi-Strauss declared a new "mathematics of man" based on a structuralist conception of mathematics, a mathematics, Lévi-Strauss argued, which would make mathematics ever more relevant to the study of society, Kuhn's structuralist approach to history effectively scissored off the history of mathematics from the history of science (and society).[36] Despite Birkhoff's assertion, mathematicians had no problem understanding historians' abstractions; they had other reasons for their antagonism to Kuhnian ideas. As Peter Gordon remarks, Kuhn's *Structure* and structuralism shared a "common basic model of historical change as effecting sudden ruptures or breaks in the rules of intelligibility."[37] In the following years, historians' belief in the clean breaks between historical periods only increased: "Ironically, a discipline that devoted itself to the explanation of change over time incorporated the idea of radical discontinuity into the arsenal of its own 'normal science.' "[38] The structuralist conception of mathematics similarly called for radical discontinuities in mathematical theories. However, it was precisely this tendency to construct sharp breaks in mathematical theory that turned mathematicians toward the history of mathematics in search of mathematical meaning. Mathematicians probed the historical record in search of *continuity* in mathematical ideas, not radical breaks. The whole point of studying history, according to them, was to highlight its continuities.

Historians of mathematics, finding historians of science less receptive to their work, could have turned back to the history of ideas. In the 1940s and 1950s, the history of science was an integral part of the history of ideas. It was only with the professionalization of the field, and in part as a response to Kuhn's work, that the two went separate ways.[39] The history of ideas was an adequate home for the history of mathematics, as mathematics for the most part is a textual activity, and the history of mathematics requires the sort of close textual analysis historians of ideas specialize in. However, like the history of science, the history of ideas was in the midst of its own revolution.

From History of Science to History of Ideas

While Kuhn criticized the logical positivists' belief in the existence of a pure sense-data language, as Galison points out, he shared their core assumption "that the activity of science is principally to be understood as an unraveling of the difficulties of language and reference."[40] Kuhn might have turned philosophers' attention to scientific activity as pursued, as opposed to idealized, but he continued to focus on the linguistic aspects of science as the locus of philosophical investigation. In the fifth chapter of *Structure*, "The Priority of Paradigms," Kuhn appeals to Wittgenstein's *Philosophical Investigations* in order to explain how, despite the lack of explicit rules, paradigms can be recognized.[41] Kuhn explains that Wittgenstein asked "what we need to know . . . in order that we apply terms like 'chair,' or 'leaf,' or 'game' unequivocally and without provoking argument."[42] Kuhn then goes on to explain how Wittgenstein rejected the idea that we can use a certain word, such as "game," because there is a set of characteristics or attributes that are common across all games. Instead, Wittgenstein forwarded the notion of "family resemblance," explaining that we can apply the word "game" to a new activity we have not encountered before because we recognize that it bears some resemblance to other activities that we have previously called games. "Something of the same sort may very well hold for various research problems and techniques that arise within a single normal-scientific tradition," Kuhn explains.[43] Kuhn's point was that during a period of normal science there was no clear set of rules that defined it; rather, the research problem and techniques shared some resemblance that scientists learned to recognize through their training and education. Wittgenstein's insight about the function of language provided Kuhn with a theoretical justification for his concept of paradigm and hence his overall theory of historical change in the sciences.

Seven years later, another historian turned to Wittgenstein for inspiration. Skinner similarly homed in on Wittgenstein's analysis of language and the idea that words did not have a coherent and stable meaning, but the lesson he took was different. Rather than a theory of historical change, for Skinner, Wittgenstein offered a guide for historical methodology. Since words did not have meaning outside of their use, the only thing a historian could do was examine how words were put into use. Skinner's interest in semantics was shared by other social scientists. Attention to the meaning of concepts defined much social scientific research in the postwar period. Moreover, the idea that political or ethical concepts did not have inherent meaning was not novel. As I noted in chapter 3, in considering the meaning of theory in the social sciences, Anatol Rapoport, for example, located the mistake of past researchers in their assumption that "entities called politics, society, power, welfare, tyranny, democracy, milieu, progress, etc., actually exist, just as cats, icebergs, coffee pots, and grains of wheat exist."[44] However, sociologists and psychologists hoped to stabilize the meaning of such concepts within a given social scientific theory. The concept of utility might not have any essential meaning, but within the theory of games it was formally established and hence could be unambiguously applied (so they argued) to the study of society. Skinner, following Wittgenstein, took a more radical approach. Since no such inherent meaning existed, all that could be done was to try and understand how concepts were put to use for particular ends at particular times by particular authors.

Skinner opened "Meaning and Understanding in the History of Ideas" by dismissing what he claimed were the two dominant methodologies in the history of ideas. The first held that the context *determined* the meaning of a given text; the second, to the contrary, insisted on the autonomy of the text (here he had formalists such as the New Critics in mind). Skinner was more sympathetic to the claims of the contextualizers, agreeing that some context was necessary to understand a text, but he vehemently opposed the idea that external factors fully *determined* a text's meaning. It was, however, his critique of those promoting decontextualized analysis that showed the force of his analysis. In their works, he identified a commitment, at times explicit but often implicit, to essentialized and universalized ideas. Skinner singled out the work of Arthur Lovejoy as typifying the problem: "The great mistake lies not merely in looking for the 'essential meaning' of the 'idea' as something that must necessarily 'remain the same,' but even in supposing that there need be any such 'essential' meaning (to which individual writers 'contribute') at all."[45] Skinner explained that such essentializing assumptions led historians to produce ahistorical analysis or, as he put it, subscribe to a series of mythologies.

Among the myths Skinner identified were: (1) a tendency for "crediting a writer with a meaning he could not have intended to convey, since that meaning was not available to him," (2) assuming that an idea existed in some ahistorical realm just waiting to be discovered, (3) searching for anticipators of an idea, and (4) endlessly debating "whether a given idea [might] be said to have 'really emerged' at a given time, and whether it [was] 'really there' in the work of some writer."[46] Skinner reasoned that, consciously or not, historians of ideas felt justified in applying their contemporary understanding of an idea to a reading of historical texts because they fundamentally held that the idea itself was timeless. Historians' "philosophical mistakes" engendered their ahistorical practices.

By the early 1970s, historians of mathematics and some historically minded mathematicians began to sound an almost identical critique of the history of mathematics. Kenneth May came to the history of mathematics late in his career. He earned his BA in mathematics in the 1930s from the University of California, Berkeley, where his father taught political science, and he was active in communist circles. Having traveled to the Soviet Union in 1937, May returned to Berkeley to work on his PhD in mathematics, but his work soon came to a halt when his father publicly disowned him for his Communist sympathies. As a result, the university's Board of Regents revoked his teaching assistantship. When World War II broke out, May enlisted in the army, only returning to Berkeley at the war's end to complete his PhD thesis, "On the Mathematical Theory of Employment."[47] He was one of the mathematicians who helped usher axiomatic thinking into the postwar social sciences. During the 1950s, May began publishing articles about mathematical teaching and book reviews of histories of mathematics.[48]

In 1966, May moved to the University of Toronto, where he began working on a long-term project to produce comprehensive general reference tools for the history of mathematics. In 1973, he published *Bibliography and Research Manual of the History of Mathematics*, a comprehensive account of the literature in the field classified across period, topic, and country.[49] A year later, May became the first editor of *Historia Mathematica*, which was founded in response to growing interest in the field. The journal emerged from an international effort. It was first conceived during the 1968 Twelfth International Congress on the History of Science in Paris and in 1971, as part of the founding of the International Commission on the History of Mathematics.[50] While much thought and work was devoted to establishing *Historia Mathematica*, questions as to what the history of mathematics was and for whom it was written were not foreclosed by the journal's publication, but rather thrown into sharp relief. In the second issue of the journal, May published "Should We Be Mathematicians, Historians of Science, Historians, or Generalists?"[51] The answer, according to May, was all of the above.

May was uniquely positioned to recognize the difficulty in situating the history of mathematics simultaneously in history and mathematics. In 1975, May published a two-part commentary in the journal under the headline "Historical Vices." The first historical vice historians of mathematics were guilty of revolved around "logical attribution":

X knew A

B is a logical consequence of A

Therefore, X knew B.[52]

It was surprising, May noted, how often a writer would use their mathematical skills to derive one result from another and then attribute this later result to the work of a past mathematician. Such logical schemas took different forms, according to May.[53] In all instances, he explained, the writer was making assumptions about mathematical knowledge in the past based on mathematical knowledge in the present. The second vice May identified was the constant pursuit of priority, or rather, who proved it first. Mathematicians might be guided by the desire to find new results, but they should not project their concerns onto past mathematicians.[54] May's logical derivation almost perfectly encapsulated Skinner's criticism. The search for the true "birth" of an idea, or the attribution of contemporary standards to the historical past, were exactly the sort of vices Skinner identified.

Moreover, they both recognized the same culprit. "These vices arise," May concluded, "from looking at mathematics as a timeless, static, logical structure gradually revealed to man, instead of as a historically evolving human phenomenon."[55] May articulated exactly why developing the history of mathematics as part of both mathematics and history was problematic. The difficulty arose not out of misunderstanding what history was, as the conference participants hinted; it was due to two conflicting temporal understandings of what mathematics was as a body of knowledge. This point bears repeating—the disagreement was not about what history was, but what mathematics was.

That same year, the tension between the historians and the mathematicians came to a breaking point with the publication of Sabetai Unguru's "On the Need to Rewrite the History of Greek Mathematics."[56] It is not clear whether Unguru read Skinner's 1969 "Meaning and Understanding in the History of Ideas," but the two polemics bear strong similarities, both in their criticisms of contemporary historical practice and its cause. Unguru contended that historians of mathematics, often being mathematicians themselves, had adopted the view that in order to write the history of mathematics, one must first be familiar with *modern* mathematics. This was taken as a truism whether one was writing about the history of ancient mathematics or the history of eighteenth-century mathematics: "Whig history *is* **history** in the domain of the history of mathematics; indeed, it is still, largely speaking, the standard, acceptable, respectable, 'normal' kind of history, continuing to appear in professional journals and scholarly monographs."[57] The problem was the same one Skinner had identified—in bringing their own contemporaneous understanding of the field to bear on the historical past, mathematicians were quick to attribute to historical mathematicians ideas they simply could not have had.

When it came to Greek mathematics, this ahistorical approach, Unguru explained, gave rise to the concept of "geometric algebra"—the idea that even though the Greeks presented their ideas geometrically, through figures, their thinking was fully algebraic. Geometric algebra denoted the presentation of algebraic ideas through geometric figures. A simple example of this is the presentation of the concept of x^2 as a square with sides of length x. Explaining the gist of that approach, Unguru wrote, "We are, then, not only authorized to look for the algebraic 'subtext' (so to speak) of any geometrical proof, but it is indeed wise (historically!) always to transcribe the geometrical content of any proposition in the symbolic language of modern algebra, especially when the former is particularly cumbersome and awkward, while the recourse to the latter always makes the logical structure of the proof clear and convincing, without thereby losing anything not only in generality but also in any possible sui generis features of ancient ways of doing things."[58] What historians of mathematics were doing, Unguru explained, was assuming that the *content* did not in any way depend on the *form* in which ideas were expressed.[59] To read Greek mathematics with a modern eye that was already trained to move freely between algebra and geometry was the "safest method for misunderstanding the character of ancient mathematics."[60] Why, Unguru asked, would the Greeks choose to present their work in such a "cumbersome" geometric way if they were truly algebraic?

Skinner asked a similar question. He noted that, besides crediting a writer with a meaning they could not have intended, historians of ideas also tended to "read in" a doctrine where it was not. "In all such cases," he wrote, "where a given writer may appear to intimate some 'doctrine'; in something that he says, we are left confronting the same essential and essentially begged question: if all the writers are claimed to have *meant* to articulate the doctrine with which they are being credited, why is it that they so signally failed to do so, so that the historian is left reconstructing their implied intention from guesses and vague hints?"[61] The dangers of essentializing and dogmatic reading were so great, according to Skinner, that the only way to guard oneself against holding such a position was to practice contextual reading: "It may well seem, however, that even if such mythologies proliferate at this level of abstraction, they will scarcely arise, or will at least be never much easier to recognize and to discount, when the historian comes to operate on the level simply of describing the internal economy and argument of some individual work."[62] This is the lesson intellectual historians took from Skinner: trying to work across texts and periods would inevitably commit the historian to an essentializing conception of ideas, and, hence, to taking part in mythology. But what if continuity were possible without all the pitfalls Skinner so rightly identified?

Mathematical History

Considering the vehemence of his writing, it is no surprise that after Unguru's paper appeared in the *Archive for History of Exact Sciences*, mathematicians hit back. Whereas May was first and foremost a member of the mathematical community, Unguru was a historian. Van der Waerden, Hans Freudenthal, and André Weil responded to Unguru in print, defending the concept of geometric algebra. Unguru's paper, they argued, was simply polemical and itself missed some necessary contextualization. Their response revolved around the meaning of algebra, what counted as symbolic thinking, and what did and did not count as historical evidence. Beyond these technicalities, however, lurked a deeper disagreement. Unguru's criticism, as noted earlier, was not limited to the history of ancient mathematics. It questioned in principle the use of modern mathematical knowledge in the analysis of historical work. Mathematicians, however, did not acknowledge this charge. For example, in his reply, Freudenthal explained that words such as "line," "square," and "product" had both geometric and algebraic connotations for the Greeks. "When applying the theorem one is allowed to disregard their geometrical connotation, which in fact can only hamper the progress of thought," Freudenthal explained. "This is quite a normal feature of mathematics."[63] But this was exactly what Unguru was writing against. Unguru protested the very notion that one could speak of a "normal feature of mathematics" that transcended historical period.

Faced with a strong backlash by the mathematicians, Unguru wished to publish a response in the *Archive*, but he was refused, so he turned instead to *Isis*. His response reiterated many of his earlier points: "If scholars continue to neglect the peculiar specificities of a given mathematical culture, whether as a result of explicitly stated or implicitly taken-for-granted assumptions, then by definition their work is ahistorical and should be recognized as such by the community of historians."[64] It also highlighted that the disagreement was less about history and more about the nature of mathematics. "History," Unguru insisted, "is the attempt at understanding each past event in its own right."[65] History was the study of events of the past that paid particular attention to change and the particular or idiosyncratic. Echoing Skinner, Unguru concluded that the historian should not focus on what ideas in the past and the present had in common but, to the contrary, what set them apart. "I shall not presume to define here what mathematics is," Unguru wrote. "But," he added, "I can say with safety what mathematics is not. It is certainly not history."[66] But what if it was? Or, at least, what if practicing mathematicians themselves approached mathematics as historical?[67]

Clifford Truesdell was another mathematician who in the postwar moment turned to history, but unlike some of his colleagues, he posed his historical studies in opposition to those of the historians. Truesdell's research was wide-ranging, encompassing everything from continuum mechanics and thermodynamics to historical research on the work of Leonhard Euler and Leonardo da Vinci. Truesdell, who was a founding member of the journal *Archive for History of Exact Sciences*, published in 1968 *Essays on the History of Mechanics*, which collected several lectures Truesdell had given in previous decades. "If these lectures find any favor with professional historians of science," Truesdell wrote in the introduction, "I shall be humbly thankful for their toleration of a book not intended for them."[68] Truesdell posed his historical studies as defiantly distinct from those of historians of science.

His stance was surely motivated by professional antagonism, and he was quick to blame historians for lacking the necessary technical knowledge to understand the work of the past. However, more fundamentally, his objectives differed from those of historians. The problem with historians of science, he explained, was that they approached their subject matter as dead. "Such historians," he wrote, "remind me of those taxonomers, perhaps of only fabulous existence, who cannot recognize a particular plant unless they see a sprig of it dead, dried, and pasted to a sheet of paper."[69] For him, science (in the past or present) was alive, and served as a source of inspiration for the present. The history of mathematics was a subject pursued by mathematicians because it was a lively history that said as much about how things *were* as how things *are* and how they someday *could be*. I take this claim seriously. For mathematicians, sixteenth-century mathematics could speak to the present. This is not to say that mathematicians do not uphold some idea of progress, but progress, for these mathematicians, does not jettison the past. Instead, it continuously enfolds it into the present. This lively history does not just impact the way they approach the history of mathematics, but how they look at mathematics itself as a research enterprise and how they conceive of their own work.[70]

The physicist John Wheeler recalled that Hermann Weyl once arranged to give a course on the history of mathematics at Princeton University. "He explained to me one day," Wheeler wrote, "that it was for him an *absolute necessity* to review, by lecturing his subject of concern in all its length and breadth. Only so, he remarked, could he see the great lacunae, the places where deeper understanding is needed, where work should focus."[71] Not every working mathematician or fresh PhD held such a rich and totalizing view of mathematics, but for many working mathematicians, especially those concerned with mathematics as a whole, history said as much about the present as it did about the past. In the 1974 conference proceedings, Morris Kline referred to Weyl's "statement that perhaps mathematics [was] no more than the history of mathematics."[72] Whether Kline was paraphrasing or directly quoting is immaterial. If mathematics is the sum total of its history, then it is no surprise that mathematicians felt protective of it.

In 1979, Weil returned to the subject during an address to the International Congress of Mathematicians in Helsinki. In "History of Mathematics: Why and How," Weil explained that the history of mathematics was important because it served as a rich source of inspiration. "Eisenstein," Weil noted, "fell in love with mathematics at an early age by reading Euler and Lagrange."[73] If future physicists showed their inclinations by tinkering with technologies, taking apart radios and putting them back together, then future mathematicians disclosed their interests by reading mathematicians of the past. History was not just a gateway to a mathematical career, but an integral feature of mathematical activity. "No doubt," Weil said, "a young man can now seek models and inspiration in the work of his contemporaries; but this will soon prove to be a severe limitation. On the other hand, if he wishes to go much further back, he may find himself in need of some guidance; it is the function of the historian, or at any rate of the mathematicians with a sense for history, to provide it."[74] It is revealing that Weil, one of the founding members of Bourbaki, who promoted the high modernist structuralist approach to pure mathematics, held that a young mathematician coming into the field could not be satisfied by simply studying the work of his contemporaries. Bourbaki was originally motivated by a desire to rewrite classical mathematical textbooks in a modern spirit. In their *formal* presentation, Bourbaki's texts announced a rupture, but as Weil's remarks make clear, the goal was never to completely sever mathematics from its past. Throughout his talk, Weil referred not to the history of mathematics, but to mathematical history. The distinction was meaningful, as it pointed to two separate activities.

Analyzing Weil's speech, David E. Rowe has argued that Weil's approach to history is rooted in Platonism: "One can present the development of mathematical ideas as a steadily unfolding search for Platonic truths that transcend the particular cultural context in which these ideas arose, but only by discounting the rich variety of meanings that accompanied this work."[75] This is undoubtedly true. The belief that mathematical ideas are context-independent necessitates Platonist leanings. However, debates about the history of mathematics were not, at their core, about whether a Platonist position was sustainable. Rather, they were about the source of mathematical *meaning*. The work of mathematics was on the table, not mathematical philosophy. The debates were about what it meant *to be* a mathematician and to pursue mathematical research. This is the only way to explain the fervor with which mathematicians approached the history of the field.

This becomes evident in Weil's explication of what counts as a mathematical idea and a "mathematical strategy." For Weil, *to be a mathematician* required the ability to recognize mathematical ideas in the course of their historical evolution and conceive of mathematics as an ongoing project. To be a mathematician was to be engaged in continuous conversation with the work of the past: "Mathematical strategy is concerned with long-range objectives; it requires a deep understanding of broad trends and of the evolution of ideas over long periods."[76] History was how mathematicians learned to distinguish between trivial and fundamental problems, and detect potential paths for solving them. Historical sensitivity, in other words, was indispensable to mathematical know-how. A mathematical idea, Weil conceded, could not be easily defined. A mathematician, however, could easily recognize one. While Weil did not say so explicitly, he suggested that history offered mathematicians a way of homing in on such distinctions. History was just one more tool in the hands of working mathematicians.

Throughout his talk, Weil distinguished between what he considered to be cultural and mathematical questions. He did not dismiss out of hand the work of the historian; rather, he emphasized that the historian's interests were not those of the mathematician. "The historian," he said, "tends to direct his attention to a more distant past and to a greater variety of cultures; in such studies, the mathematician may find little profit other than the aesthetic satisfaction to be derived from them and the pleasures of vicarious discovery."[77] On the other hand, he said, "The mathematician tends to do his reading *with a purpose*, or at least with the hope that some fruitful suggestion will emerge from it."[78] Taking Weil at his word, the history of mathematics and mathematical history are two distinct disciplines, a point which Unguru himself endorsed.[79]

In the concluding section of his paper, Skinner reflected on the implications of his criticism for the study of the history of ideas. He emphatically denied that the history of ideas could be justified by claiming that the classic text spoke to "perennial problems" and "universal truths."[80] Not only were the classic texts not dealing with the same problems historians in the present were facing, but there was also nothing we can learn from their answers. The historical and cultural difference was so vast, Skinner held, that it could "hardly be useful even to go on thinking of the relevant question as being 'the same' in the required sense after all. More crudely: we must learn to do our own thinking ourselves."[81] Here, then, was one important difference. For practicing mathematicians, it was not so much that there were perennial questions in mathematics, but that the historicity of mathematics itself was perennial, including a past which never stopped speaking to the present. This historicity can indicate, but does not necessitate, a Platonic position. Understood as part of mathematical activity, of what constitutes mathematical research, the continuity of mathematical ideas is not an ontological claim, but a practice-based accomplishment. It is a continuity that mathematicians enact, as opposed to claim, by looking to past work for a better understanding of their field.

When Kuhn's collection of essays *The Essential Tension* appeared in 1977, historian Christopher Berry took the opportunity to reflect on the impact of Kuhn's work on the history of ideas, suggesting that it was exactly the issue of historicity that limited the applicability of Kuhn's work outside of science. "Hannah Arendt in *The Human Condition*," Berry wrote, "can premise an acute critique of contemporary society on a reading of Aristotle, but nothing similar can be done with Aristotelean vis-à-vis Einsteinian physics. It is not just that the past is always there, but that no currently dominant 'reading' of it can preclude its future utilization."[82] Berry was thinking about the history of political and philosophical ideas, but his analysis held just as true for mathematics. "It is not just that the past is always there" but, as Dieudonné noted, once neglected a mathematical idea can always find new life. All that is necessary is a mathematician who will adopt it as such. What the historical work of mathematicians points to is less Platonism than continuous engagement.

A Weekday Platonist

Since Skinner has so thoroughly critiqued the possibility of a history of an idea as such, intellectual historians have, for the most part, shied away from temporally vast studies. The philosophical mistakes Skinner identified have haunted intellectual history ever since, as is evident in more recent claims to expand beyond close contextual studies. More than four decades after the publication of Skinner's paper, David Armitage has called for a *longue durée* intellectual history that will not shy away from examining the "deep pasts" of "central concepts in our political, ethical and scientific vocabularies."[83] However, Armitage is quick to explain that "no intellectual historian would now use Lovejoy's creaking metaphors of 'unit-ideas' as chemical elements, nor would they assume that the biography of an idea could be written as if it had a quasi-biological continuity and identity through time, along with a life cycle longer than that of any mortal human subject."[84] Armitage asserts that, once reinvigorated, the history of ideas, or, as he insists, a history *in* ideas, would fundamentally already incorporate into itself the forceful critique of Lovejoy's approach, and any resemblance between the old and new method would be "artificial."

Taking the opportunity to consider the history of ideas in light of Armitage's call for a return, Martin Jay begins by asking what is "the nature of the beast whose longterm history is being tracked, understood variously as ideas, concepts, and metaphors." Jay notes that the idea of an "idea" is already unstable, and emphasizes that the old history of ideas had an expansive understanding of what constituted an idea, recognizing both its connotative and denotative aspects. However, he too, denounces Lovejoy's essentialist ontology: "The main impulse behind the Lovejoyian approach was to isolate manifest ideas or latent unit-ideas from their social, psychological, and material contexts of origin and reception, and seek to capture their core meaning, a meaning that endured over time despite the vicissitudes of the ideas' development."[85] The question is the same one Mehrtens raised: What is the "content" of an idea outside its diverse set of manifestations?

Even more sympathetic to Lovejoy, Darrin McMahon has called for a return to history of ideas. As he notes, Skinner's and J. G. A. Pocock's criticism of Lovejoy was exaggerated. McMahon notes that Lovejoy himself rejected the idea that a text could be understood on its own terms, and was critical of New Criticism's strict formalist approach. On the contrary, as McMahon explains, Lovejoy insisted on a capacious reading of a text that transcended disciplinary boundaries moving between philosophy, theology, art, and science. The Cambridge School and Lovejoy's project, McMahon writes, "are more similar than they have acknowledged."[86] However, he, too, emphasizes that Skinner's critique of Lovejoy's "unit-ideas" was both justified and important: "Skinner and Pocock's pointed reminder that ideas do not somehow magically exist outside of the languages in which they are embedded and the uses to which they are put is an insight that few practicing intellectual historians would choose now to discount."[87] In other words, one must guard above all against Platonism.

To protect against such philosophical fallacies and nonetheless offer a *long durée* intellectual history, historians have forwarded several methodologies. Armitage terms his solution "serial contextualization." By adopting both a diachronic and a synchronic approach, historians, according to Armitage, can offer a transtemporal history *in* ideas. Calling for a return to the history of ideas within the history of science, John Tresch instead emphasizes the materialization of ideas. Far from existing in some ambient ether, Tresch argues for the study of ideas that are "enacted, embodied, elaborated, and contested in concrete settings, institutions, representations, instruments, and practices."[88] Jay, for his part, points to conceptual history in the fashion of Reinhart Koselleck, insisting on the "desubstantialization" of concepts as a more "promising approach."[89] Conceptual history, as Jay writes, similarly posits no stable meaning for a concept. Indeed, as Koselleck puts it, "A concept can possess clarity, but must be ambiguous."[90] Such conceptual history is inherently big, as it by necessity includes a diverse set of meanings often quite different from contemporary ones.

All these suggestions, however, do not solve the philosophical quandary that Skinner so clearly posed. There is no escaping it: What unites such broad-ranging accounts? If it is an inherent instability of meaning that makes a concept a concept, is it even an identifiable referent? Expanding on the approach of conceptual history, Jay writes, "Although not based on the policing of lexical boundaries via normative definitions, concepts appear to operate through the logic of subsumption or at least the seeking of common denominators underlying different usages."[91] Another way to say it is that in order to be a meaningful unit of historical analysis, a concept or an idea, despite its inherently changing nature, must have some "common denominator" that can be recognized as such. But what is this core? How do we recognize it?

Skinner had already acknowledged the problem in his 1969 essay, but in 2002 he returned to it in *Visions of Politics*, and did so, not surprisingly, via a comparison with Koselleck.[92] Skinner writes that, contrary to what others have claimed, he does not see his approach in any way conflicting with that of Koselleck. "Koselleck and I both," Skinner writes, "assume that we need to treat our normative concepts less as statements about the world than as tools and weapons of ideological debate."[93] Skinner explains that he does not object to writing conceptual histories, at least not ones that seek to uncover how concepts have been mobilized over time. For Skinner, much is at stake in studying conceptual histories. As he explains, determining how to use normative terms is one of the ways we make sense of the world around us, and thus changing how we apply our evaluative concepts is "one of the engines of social change."[94]

Having established that the "phenomenon of conceptual change" is important in historical studies, Skinner surprisingly adds, "My almost paradoxical contention is that the various transformations we can hope to chart will not strictly speaking be changes in concepts at all. They will be transformations in the applications of the terms by which our concepts are expressed."[95] How can Skinner argue that there is nothing beneath and behind the *uses* of an idea and simultaneously claim that the concept itself does not really change? It seems obvious that he means more than just linguistic stability (after all, concepts and words are not equivalent). Skinner offers as an example his notion of rhetorical redescription, whereby the same term can take on a completely different and even oppositional meaning. In *The Prince*, for example, Machiavelli turns on its head the virtue of liberality by insisting that the behavior that is often described as liberality should in fact be understood as ostentatiousness. Rather than a virtue to be upheld and celebrated, in Machiavelli's hands, liberality becomes a vice. The question to be answered, then, is whether such a process indicates a conceptual change or not. Does the concept of liberality truly change? Or is it just that the possible situations in which the concept can be appropriately applied are restricted?

This seems to be the crux of the issue. On the one hand, there are those who will argue that what is involved is precisely not conceptual change. What liberality means remains the same. The only thing that changes is that in the aftermath of Machiavelli's writing, certain actions of the prince, which in the past could have been described as exemplifying said virtue, can no longer be described as such. Since for Skinner, the meaning of concepts is made through their use, this is not where he ends. For him, rhetorical redescriptions do account for conceptual change.

> This caution strikes me as correct and important, but the fact remains that the outcome of such debates will nevertheless be a form of conceptual change. The more we succeed in persuading people that a given evaluative term applies in circumstances in which they may never have thought of applying it, the more broadly and inclusively we shall persuade them to employ the given term in the appraisal of social and political life. The change that will eventually result is that the underlying concept will acquire a new prominence and a new salience in the moral arguments of the society concerned.[96]

Mathematicians and historians of mathematics struggle with a similar conundrum. Did the concept of a number change with the acceptance of imaginary numbers or did mathematicians simply have different concepts of what constituted a number? For them, the stakes are not moral but epistemological. This, after all, was the problem that Frege posed in his *Habilitation*: what happened to the concept of "quantity" in light of the discovery of imaginary numbers?

I point to the history of mathematics because it is in mathematics alone that Platonism is an acceptable position. Reuben Hersch claimed, "The typical working mathematician is a Platonist on weekdays and a formalist on Sundays."[97] The Platonist position is not justified on philosophical grounds; rather, it is a characteristic of mathematical research. And while in the hands of historically minded mathematicians (and some historians) Platonism often leads to anachronism and a repetition of many of the so-called myths Skinner identified, I think that the fervor with which mathematicians turn to history can also point elsewhere. After all, if all they were after was an abstract definition of a given concept or idea, they could have found that in their textbooks. There was no need to look to the past. With the danger of reading too much into their work, what mathematicians seem to suggest is that a given concept *is* the sum of its history. The concept of a line is *simultaneously* a Euclidean line, a geodesic in a Riemannian manifold, the implicitly defined concepts that satisfy Hilbert's axioms of geometry, and the sum total of the interactions between them. In Skinner's critique several historical offenses are grouped together: the danger of presentism, an unfounded belief in timeless ideas, and a commitment to progress. Their combined dangers scared intellectual historians away from broad historical analysis that transcended historical time and place. However, this grouping is not inevitable. Even in mathematics one can have continuity without either Platonism or progressivism.

Daniel Wickberg has argued that the critiques of Lovejoy, both in his time and following Skinner, have simply mischaracterized (or misunderstood) his project. Lovejoy, Wickberg writes, "was neither a radical nominalist nor a Platonic universalist; for him, the question [of] how much remained consistent in an idea was a historical question."[98] Instead Wickberg locates the difference between Lovejoy and later historians of ideas in the former's belief that ideas might have a force of their own that acts upon the writer of a text. Whereas Skinner and his colleagues were concerned only with how ideas were used by writers, Lovejoy held that "Ideas ha[d] a logic of their own, that their users [were] in some sense coerced into positions and conclusions that they [were] themselves not aware of."[99] Such a position might be the closest description of what mathematicians had in mind. The key here was not Platonism, but contingency and continuity. I would venture further that mathematicians did not turn to history to offer a narrative of progress, which they had already established in their strictly technical work, but rather to understand their work as existing in conversation with the past.

Koselleck's *Begriffsgeschichte* is the anthesis of Frege's *Begriffsschrift*. Through his concept notation, Frege tried to find a symbolic language with which to fix the meaning of concepts. His goal was to define concepts to make them stable. On the other hand, conceptual histories do exactly the opposite. The goal is not to stabilize the meaning of concepts but to mine the historical record for the ways in which meaning changes over time and to further argue that any attempts at definition are futile. Furthermore, one can argue that conceptual history arose out of the failure of logistics and philosophical attempts to identify a universal language. Mathematicians, I suggest, sought to find a middle ground between the two positions. In their research, they sought to fix concepts; mathematical objects had to be clearly defined. To do so, they did not appeal to some universal language of logic, but to a given axiomatic system, whose assumptions and rules of inference were clearly stated. On the other hand, the meaning of the concepts and ideas they considered were not confined within a given axiomatic system. This is why mathematicians appealed to history.

Nietzsche famously claimed that "all concepts in which an entire process is semiotically concentrated elude definition; only that which has no history is definable."[100] Mathematicians, for whom definitions are bread and butter, still recognize that mathematical concepts have a history. In the postwar period, this was part and parcel with their modernist rewriting. The force of their project demonstrates that an object is never stable. A new theoretical framework can always arise that will make concrete that which has formerly been abstract. And this force applies as much to the past as it does to the future.

Epilogue

The first thing mathematician George Francis did after he received tenure at the University of Illinois at Urbana-Champaign was spend his days in the university's library reading late nineteenth-century geometry books. He left behind the research problems he had been working on until that point and instead turned his attention to the illustrations that filled these almost century-old books. It was during this time that he began to be "serious" about drawing. Francis had loved to draw pictures since he was a kid, and with a "famous painter" on his father's side and a "competent illustrator" on his mother's side, artistic talent had been running in his family for decades.[1] Still, in 1977 Francis decided to merge his passion for drawing with his love for mathematics and make it the focus of his academic career. Ten years later, he published *A Topological Picturebook*, half a drawing manual and half an illustrated guide to the field of low-dimensional topology. In the introduction, Francis reflects on the semester he spent studying the work of nineteenth-century geometers. Theirs, he writes, "was a wonderfully straightforward way of looking at rather complicated things . . . They drew pictures, built models, and wrote manuals on how to do this. And so they captured a vivid record of the mathematics of their day. I resolved to try to do the same for the mathematics of my contemporaries."[2] For Francis, the mathematical past had much to say to its present.

Francis began his doctoral studies at Harvard University. After a visit to the University of Michigan, during which he recalled having had more engaging conversations about mathematics in one weekend than he had had during three years at Harvard, he transferred. He eventually earned his PhD in 1967. He spent an additional year in Michigan, and in 1968 he and his wife moved to the University of Illinois at Urbana-Champaign. At first, because Francis was not interested in becoming an overly specialized researcher, he began publishing papers in a variety of fields, including statistics and control theory. After a senior colleague in the department informed him that a lack of clear specialization might cost him his promotion, Francis turned his full attention to a specific class of problems in low-dimensional topology that arose out of his dissertation. In 1977, he received tenure. It was at this point that he returned to drawing.

When Francis began working on his dissertation on low-dimensional topology, which studies topological space in four and three dimensions, was not a fashionable field. To be sure, there were a number of mathematicians in the US and abroad working in the field, but it was high-dimensional topology and algebraic topology that represented the lion's share of work in the field. Francis was well aware of that fact, but the decision to specialize in low-dimensional topology was a conscious one. "I was done with abstractions. I just became very concrete," he explained. Francis did not wish to start a mathematical inquiry by stating, "Let there be a space with such and such properties."[3] He was not interested in generalization. His entry point had to be a particular example out of which he could expand outward. It was this disposition that would later push him into mathematical illustration. He wished to balance algebraic logic with geometric visualization, and low-dimensional topology more easily lends itself (at least in principle) to geometric imagination and intuition than topological spaces in higher dimensions. His academic path moved in the opposite direction from that of Steenrod, who three decades earlier had made the turn away from visualizable and concrete mathematics toward the realm of abstractions.

Besides mathematical textbooks from the previous century, during his sabbatical Francis studied drawing manuals, medical atlases, and the work of famous artists such as M. C. Escher and Albrecht Dürer. In personal notebooks he maintained during the period, a shift is noticeable from his early simple line drawings, which often looked flat and suggestive, to more systematic and complex illustrations, which took advantage of perspective and shadowing (see Fig. E.1). As his drawing skills improved, Francis began illustrating his colleagues' work. When mathematician William

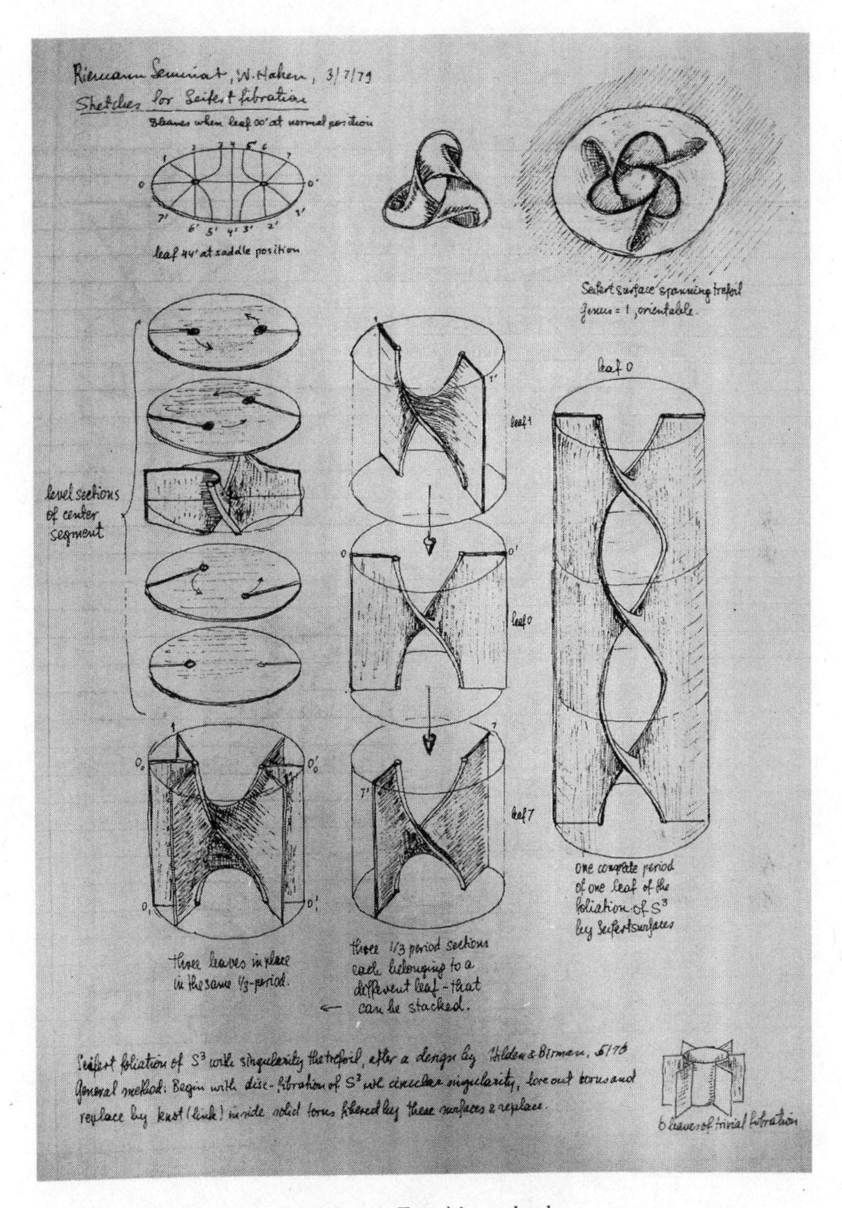

FIGURE E.1. Two illustrations from George Francis's notebooks

Source: George Francis

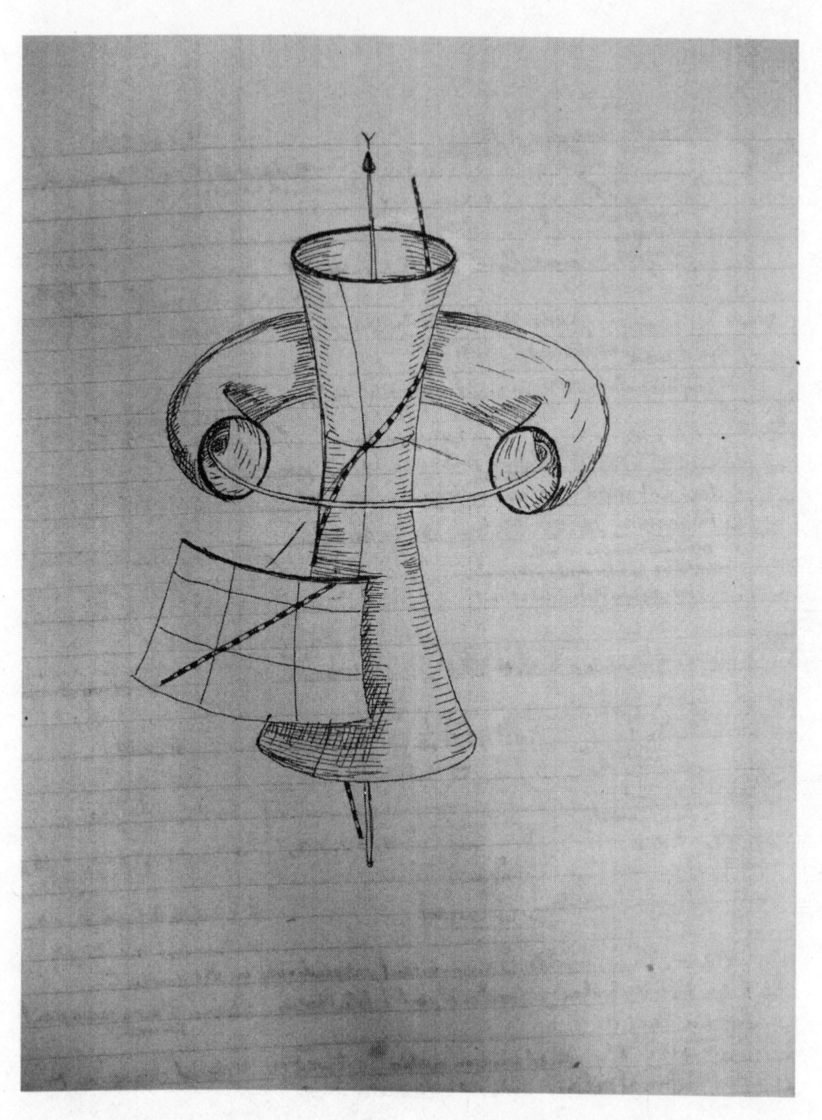

FIGURE E.1. (*Continued*)

Abikoff joined the Department of Mathematics at his university, Francis illustrated the material Abikoff presented during a seminar. Eventually, these illustrations appeared in a monograph as well as a separate article that Abikoff published, respectively, in 1980 and 1981.[4] Francis's drawings gained further traction when he began collaborating with mathematician William Thurston.

Thurston was a pioneer in the field of low-dimensional topology, and his ideas profoundly influenced the growth of the field over the following decades.[5] From the start, Thurston's approach to mathematics was unique. "Thurston's style of exposition is special," wrote one mathematician, "in that it asks the reader to participate actively in what's going on by providing room for mental images."[6] One of Thurston's later papers instructed the reader:

> Imagine walking in a barren desert when you see the space in front of you begin to shift. You are startled, and stop. You see a vertical, straight fracture where the left side does not quite match the right: the images overlap ever so slightly. At first you think your vision has gone bad, maybe you have become cross-eyed. However, when you turn your head and move from side to side, the fracture does not turn or move with your head and eyes. When you circle around at a wide distance, you see that the fracture is not fixed on the ground or on the distant scenery, but is localized on a line going straight up into the sky.[7]

The paper, Thurston explained, was not about the theory, but the "phenomenology" of 3-manifolds. What he wished was for his colleagues to see *what* he saw and *how* he saw. It is not surprising, therefore, that in the early 1980s Francis illustrated some of his mathematical work.[8]

In 1982, Thurston won the prestigious Fields Medal. The enthusiasm his work generated points to a much broader change in American mathematics. By the mid-1970s the emphasis on axiomatics and structure was on the decline. During the 1974 workshop on the history of mathematics described in the previous chapter, Felix Browder already remarked that the Bourbaki style that had dominated mathematical activity during the previous decades was decreasing in importance. "There has been a gradual but decisive change of position on the part of the core group of the most influential mathematicians towards the rejection of any sort of utopian concept that mathematics consists exclusively or even primarily of structures in the abstract sense. The programmatic proclamations of the 1950s toward such an identification, put forward, for example, in articles by Bourbaki and by Marshall Stone, would be very inappropriate today in terms of the thrust and emphasis of the major areas of mathematical research."[9] The absolute reign of high modernist emphasis on abstraction and generalization was coming to an end, as new fields and new approaches were taking center stage.[10] What explains this transformation?

The answer, in part, is reducible to numbers. During the 1950s, the American mathematical community was still relatively small. In the 1970s, older mathematicians talked with nostalgia of the old days, when mathematical meetings were intimate and most of those in attendance knew one another. Young mathematicians coming into the field in the 1950s could—through their studies, attending conferences, and reading publications in top mathematical journals—be well aware of the dominant mathematical approach and know what did and did not count as fundamental contributions to the field. By 1970, the number of mathematicians across the United States had been increasing steadily for a decade. There were more departments, more conferences, and more journals. This expansion meant that there was more space for variations from the norm. Many mathematicians took advantage of this and began challenging the idea that axiomatics was the only legitimate approach to mathematical research. The rise of more intuitive and visualizable approaches to geometry and topology exemplified this change.

There were plenty of mathematicians throughout the 1950s and 1960s whose natural inclination was toward the more intuitive and concrete aspects of geometry. The work of mathematician Donald Coxeter, whose geometry was at once tactile and visualizable, is just one example.[11] However, on the whole, such work was not placed at the center of mathematical activity. By the late 1970s, the visual aspect of geometry still had not moved to the center of mathematical activity, but more and more mathematicians argued that it should be recognized as integral to mathematical research and education. In August 1980, Jean Dieudonné, Branko Grünbaum, and Robert Osserman convened in a panel to discuss the question "Is geometry dead?" The provocative title for the panel was a response to a proclamation Dieudonné had made twenty-one years earlier during a conference on elementary education in France. There, he rose to the podium to declare "Euclid must go!" As the abstractionist approach that Bourbaki exemplified was waning, it was time to reevaluate the place of geometry in mathematics and the panel represented the changing mathematical landscape. Dieudonné stood for the abstract and algebraic study of geometry that had dominated the past three decades, while Osserman, a Stanford geometer, and Grünbaum, who helped reinvigorate mathematical interests in tiling and patterns, advocated a more capacious approach that did not seek to discredit the use of abstraction, but to balance it with more intuitive and concrete studies.

True to form, Dieudonné chose to answer the question posed by the panel by turning to history. Dieudonné argued that geometrical ideas were as fundamental then as they had been in the past. However, what

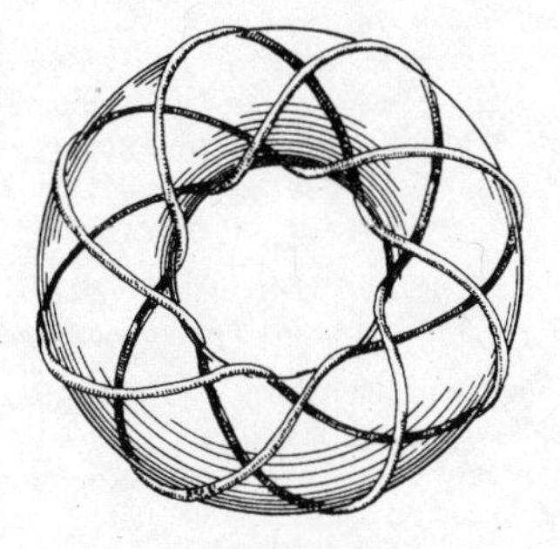

FIGURE 1. A torus knot of type (3, 8). It can be placed on a torus so that it winds 3 times around the short way while going 8 times around the long way.

With any nontrivial knot K there is associated a whole collection of other knots, known as *satellites of* K; these are knots which are obtained by a nontrivial embedding of a circle in a small solid torus neighborhood of K. Here, "nontrivial" means that the embedding is not isotopic to K itself and is not contained within a ball inside the solid torus. A knot is a *satellite knot* if it is a satellite of a nontrivial knot.

FIGURE 2. A knot and a satellite of it.

2.5. COROLLARY. *If* $K \subset S^3$ *is a knot,* $S^3 - K$ *has a geometric structure iff* K *is not a satellite knot. It has a hyperbolic structure iff, in addition,* K *is not a torus knot.*

Indeed, the complement of a knot is always prime, and the torus decomposition is nontrivial exactly when K is a satellite.

Corollary 2.5 was first conjectured by R. Riley [**Ri 1**] based on his construction of a number of beautiful examples, with the aid of the computer. His work gave a big impetus to me to prove Theorem 2.3.

In order to give the statement for closed manifolds, we need some more terminology.

FIGURE E.2. Two pages from William Thurston's "Three Dimensional Manifolds, Kleinian Groups and Hyperbolic Geometry," which included hand drawings by Francis.

Source: William Thurston, "Three Dimensional Manifolds, Kleinian Groups and Hyperbolic Geometry," *Bulletin (new series) of The American Mathematics Society* 6, no. 3, 1982.

A 3-manifold M^3 is called a *Haken* manifold if it is prime and it contains a 2-sided incompressible surface (whose boundary, if any, is on ∂M) which is not a 2-sphere. A prime 3-manifold whose boundary is not empty is always Haken (with the trivial exception of the 3-ball, which is often considered to be Haken anyway). Any prime 3-manifold whose first homology has positive rank is Haken.

2.5. THEOREM [**Th 2**]. *Conjecture* 1.1 *is true for Haken manifolds. A closed Haken manifold has a hyperbolic structure iff it is homotopically atoroidal.*

It is hard to say how general the class of Haken manifolds is. There are many closed manifolds which are Haken and many which are not. Haken manifolds can be analyzed by inductive processes, because as Haken proved [**Hak**], a Haken manifold can be cut successively along incompressible surfaces until one is left with a collection of 3-balls. The condition that a 3-manifold has an incompressible surface is useful in proving that it has a hyperbolic structure (when it does), but intuitively it really seems to have little to do with the question of existence of a hyperbolic structure.

A *link* L in a 3-manifold M^3 is a 1-dimensional compact submanifold. A 3-manifold N^3 is said to be obtained from M^3 by Dehn surgery along L if N^3 is obtained by removing a regular neighborhood of L, and gluing it back in by some new identification. The new identification is determined by choosing a diffeomorphism of the torus for each component of L. Two choices of diffeomorphisms ϕ and ψ give rise to diffeomorphic manifolds if $\psi^{-1}\phi$ extends to a diffeomorphism of a solid torus (the regular neighborhood of the component of L).

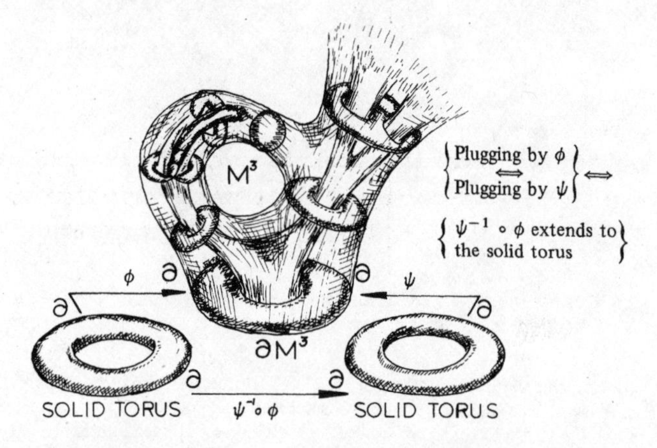

FIGURE 3. Modifying a 3-manifold by Dehn surgery. Plugging in a solid torus by φ gives a result diffeomorphic to plugging by ψ iff the diffeomorphism of the torus $\psi^{-1} \circ \varphi$ extends to a diffeomorphism of the solid torus.

FIGURE E.2. (*Continued*)

Dieudonné meant by geometry was not what his fellow panelists had in mind. When Dieudonné referred to geometry, what he was thinking of was algebraic geometry. In his view, algebra had always played an important role in geometric studies. Geometry, he concluded, had not "lost its identity"; rather, it had burst "its traditional narrow confines."[12] This was not what Osserman and Grünbaum meant by geometry. "Geometry," Grünbaum explained, "is a special way of replacing objects of the real world by 'simpler,' 'idealized' 'figures and shapes,' and then investigating the mutual relations of these. This I believe is what geometry—'intuitive geometry'—is, and should be."[13] Osserman jokingly captured the difference in their perspectives. "The answer of Dieudonné" to the question, "is geometry dead?" Osserman sarcastically declared, "would appear to be: geometry is alive and well and living in Paris under an assumed name."[14] Osserman and Grünbaum were concerned with the intuitive and concrete aspects of geometry, not its algebraic side.

Osserman, who earned his PhD from Harvard University in 1955, spent 1960–1961 as the head of the mathematical section at the Office of Naval Research. Later in his career, he became involved in efforts to make mathematics more accessible to the public. Over the years, he held public conversations on mathematics, arts, and culture with playwright Tom Stoppard, Steve Martin (Robin Williams also joined that conversation), Merce Cunningham, and Alan Alda. In his talk during the panel, Osserman conceded that the mathematical community had neglected the study of geometry in the past few decades, but he was confident that that period was coming to a close. "The primacy of the abstract, the general, and the theoretical, over the concrete, the particular, and the applied" was responsible, according to Osserman, for the decline in geometry.[15] Osserman did not object in principle to the Bourbaki approach, but he found fault with the way it had dominated mathematical activity in previous decades: "My main quarrel is with the unfortunate tendency in many circles to accept it as all of mathematics—to de-emphasize if not deny the other face of mathematics: the special, the concrete, the unstructured. And my concern is specifically that much of what I value in geometry falls in that category."[16] Osserman's critique was similar to the one expressed by Richard Courant in the 1950s. The difference was that Osserman was not calling for more applied mathematical research; rather, what he and like-minded mathematicians wanted was to make room in pure mathematics for more concrete and geometric investigations.

During his talk, Osserman offered Thurston's work as "a sign that the low ebb in the fortunes of geometry ha[d] passed."[17] When Thurston was honored with a Fields Medal two years later, Osserman was not alone in believing that geometry was experiencing something of a revival. The mathematical community's recognition of Thurston's work with the most prestigious mathematical prize also signaled to other geometrically minded mathematicians that they could fruitfully pursue their interests. The rise in visualizable mathematics was aided in no small part by the parallel growth of computer graphics. Not every mathematician had Francis's drawing skills, but many now had access to computers and recognized that they offered a new and powerful tool with which to approach topological and geometric problems. Thurston was an important figure driving research in this area as well. In the 1980s he teamed up with mathematicians and computer scientists to advocate for the creation of a computer center dedicated to the study of geometric problems.[18] The "Geometry Supercomputer Project" opened at the University of Minnesota in 1989. In an article in *Science News* announcing the project, mathematician Benoît Mandelbrot said, "This project establishes for good among mathematicians the realization that the computer is an extraordinarily useful tool for exploring geometrical problems and making conjectures, and for communicating intuitions to other people."[19] The center was later renamed The Geometry Center and remained in operation until 1998.

The growth of visualizable mathematics offered an alternative to the abstractionist and universalist tendencies of postwar mathematics by emphasizing concrete geometric investigations. It also challenged the postwar mathematical order and the division between pure and applied mathematics. One of the earliest mathematicians to turn to computer graphics was Thomas Banchoff. Like Francis, Banchoff was interested in the geometric aspects of mathematics. He built models and drew images throughout his studies, but it was his introduction to computer graphics that cemented his visualizable approach to mathematics. When he arrived at Brown University in the late 1960s, Banchoff began collaborating with computer scientist Charles Strauss to create computer graphics animation of higher-dimensional geometric surfaces.[20] Throughout the 1970s, the two produced and distributed several films that used elementary vector graphics to display projections of dimensional spaces.

In 1985, during a symposium entitled "The Merging of Disciplines: New Directions in Pure, Applied, and Computational Mathematics," Banchoff explained that he had begun forming an opinion on the nature of applied mathematics as an undergraduate. His roommate at the time spent his time in "*physics* and chemistry *labs*," while Banchoff took courses in philosophy and literature. When both men ended up pursuing PhDs at Berkeley in 1960, the differences only increased:

> Whereas I took the geometry and topology option, he chose to study *differential equations*. He began to spend more and more time with numerical computations using computers, and he would rail against the evils of bugs and batch processing. Ultimately he moved over the line into theoretical physics, where he worked in a *laboratory* on *other people's problems*. He wrote *joint papers* with *federal funding*. While I went back to Notre Dame to teach in Arnold Ross's summer program, he worked as a consultant and he began to make money. All these italicized characteristics I decided were the marks of an applied mathematician. Little did I suspect that virtually all of them would gradually begin to describe the work of pure mathematicians as well, precisely under the influence of several of the new directions in applied and computational mathematics.[21]

With the introduction of computers into mathematical research, some of the boundaries between pure and applied mathematics were no longer easy to maintain. The distinction between the two was not erased—after all, it is alive and well to this day, but there was now more room in the middle.

Finally, the rebirth of mathematical visualization also reclaimed the much longer connection between mathematics and art that was focused on patterns, form, perspective, and symmetry. Banchoff, for example, was summoned on numerous occasions to Salvador Dali's New York City hotel room to show him his newest films. Francis, who also turned to computers in the late 1980s, collaborated with several artists over the following years on exhibitions. And in the early 2000s, Thurston collaborated with Japanese fashion designer Dai Fujiwara of Issey Miyake on their 2010 collection. These collaborations reclaimed the aesthetic qualities of mathematics in the visible form rather than in axiomatic abstractions.

Osserman began his 1980 speech in response to the question "Is geometry dead?" with a discussion not of mathematics but of structuralism. What is the relation, Osserman asked, between the structuralist approach and the Bourbaki one? Osserman noted that structuralism was hard to define, but suggested that there were some typical features that united most structuralist studies: "a tendency toward the abstract, the general, and the theoretical" as well as a search for "structures" from which the subject could be approached. "A structuralist approach to a given subject," he said, "generally involved an attempt to organize the subject on a large scale, to provide a broad theoretical framework, a scientific foundation, and if possible to mathematize it."[22] Osserman called upon structuralism not to show that it shared some of the characteristics of the Bourbaki approach to mathematics, which, he wrote, "seem[ed] clear enough superficially." Rather, he asked after structuralism's collapse. Quoting Roland Barthes on literature, Lévi-Strauss on anthropology, and Julia Kristeva on the theater, Osserman wrote that what structuralists "found when they reached the end of the road was that, just as with Gertrude Stein's *Oakland*, there was no 'there' there: the original object of study seemed to fade away in the process of analyzing it."[23] The crisis of referentiality that emerged at the turn of the twentieth century led scholars away from considering anything that might be construed as a "naive approach to the subject." However, in the process, the subject matter itself had been erased.

Could the same be said about mathematics? Did mathematics experience a collapse of subject matter as well? The answer, Osserman suggested, was partially yes: "If what we mean by 'mathematizing' is abstracting the purely axiomatic and deductive aspects, and avoiding the more human activities of creating, learning, and teaching mathematics."[24] Osserman's analysis made it clear that the question that animated both the rise of high modernism and its eventual decline was "What is the source of meaning in mathematics?" This was the question mathematicians were struggling with throughout the postwar period, and as Osserman made clear, they were not alone. In the humanities and the social sciences, scholars were thinking through similar concerns.

In recovering the centrality of abstraction to the modernist project, I bring mathematics back into the conversation about midcentury American intellectual thought. One cannot understand the intellectual history of modernism without attending to mathematics. However, this fact cannot be appreciated as long as historians approach mathematics as a static and unchanging body of knowledge, or if we simply equate it with quantification. To understand the role of mathematics, it is necessary to first understand mathematics and its history; abstraction and axiomatics offer one entry.

Acknowledgments

A formal axiomatic system should be complete, consistent, and independent. While I have done my best to make this book both complete and consistent, it could never have fulfilled the criterion of independence. In writing this work and bringing it to publication, I have depended on the generosity of numerous individuals. It is a great pleasure to get to thank them. I began thinking about the history of mathematics while a student in MIT's History, Anthropology, and Science, Technology, and Society Program. There I was lucky enough to work with David Kaiser, whose wit, thoughtfulness, and support remained a constant in my academic life long after I left MIT. I could not have asked for a better role model for what academic mentorship should be like than Dave. Christopher Capozzola has been incredibly generous with his time and advice ever since he agreed to serve on my dissertation committee many years ago. Chris has pushed me outside my comfort zone and challenged me to think about how my work can speak to a broader audience. Outside of MIT, Joan Richards's enthusiasm for the history of mathematics was infectious. She has been a valuable sounding board for years and I am thankful for our numerous conversations about history, academia, and writing.

Many librarians and archivists have made this project possible. I thank Tim Engels and the staff at Brown University's John Jay Library, the Harvard University Archives, the Briscoe Center for American History at the University of Texas at Austin, Katie Ehrlich at the New York University

Archives, the University of Chicago Archives, the Archives of the American Philosophical Society, the Manuscript Division at the Library of Congress, and the National Archives at College Park, Maryland. Over the years, I have relied on the willingness of mathematicians to share their histories with me. Many were kind enough to open their homes to me. While much of these conversations did not make it directly into this book, they were foundational to how I approach this history and informed my analysis throughout. My gratitude to Thomas Banchoff, George Francis, Nelson Max, Anthony Phillips, and Charles Pugh.

For giving me the time to reconceptualize my dissertation project and to think about it anew, I am grateful to Harvard University's Society of Fellows. It is rare in academic life to have the time to think, research, and write without the pressure of time. I am one of the lucky ones, and I am grateful for the opportunity to discuss the project as it developed in such a rich collegial environment. Diana Morse, Kelly Katz, Yesim Erdmann, and Ana Novak were tireless in their support. The William F. Milton Fund enabled me to extend my research and redefine its scope. I thank the senior fellows for sustaining the most wonderful intellectual environment a young scholar can dream of. For lively and eye-opening conversations, I thank especially Barry Mazur, Noah Feldman, Peter Galison, Joseph Koerner, and Walter Gilbert. For making me a true believer in interdisciplinarity, I thank Julien Ayroles, Alexander Bevilacqua, Stephanie Dick, Len Gutkin, Daniel Hochbaum, Daniel Jütte, Abhishek Kaicker, Florian Klinger, Marika T. Knowles, Ya-Wen Lei, Jed Lewinsohn, Adam Mestyan, Rohan Murty, Christopher Rogan, and Daniel Williams.

For making academia feel less alienating and for numerous conversations along the way, I am grateful to Marc Aidinoff, Michael Barany, Dan Bouk, Moon Duchin, Clare Kim, Lisa Messeri, Christopher Phillips, Michael Rossi, Gili Vidan, Emily Wanderer, and Rebecca Woods. My utmost thanks to Stefan Helmreich and Heather Paxson for their friendship and support, and for providing a model of what an intellectual community at its best can be. Theodora Vardouli and Daniel Cardoso Llach organized a conference panel and later a wonderful conference at which I was lucky to present my work. I have been inspired by their ability to open up the space for productive and fascinating conversations among artists, historians of architecture, historians of science, and media theorists. I have presented material from this book in several conferences; for lively conversation and insightful commentary, I thank Noam Andrews, Lorraine Daston, Isabel Gabel, Megan McNamee, Raviel Netz, Michael Rossi, and John Tresch.

Volker Remmert invited me to present my work at the University of Wuppertal; members of the Center for Science and Technology Studies there asked tough questions and were kind enough not to correct my German pronunciations.

Columbia University's Department of History provided me with an academic home as I revised the manuscript for publication. Patricia Morel and Lawino S. Lurum made the transition into a new department feel seamless. When Anders Stephanson began arguing with me about the Cold War soon after we met, I knew Columbia's Department of History would be the most engaging intellectual community I could hope for. My chair, Adam Kosto, has been both graceful and generous in guiding the faculty during this pandemic. I presented material from the book at Columbia University's Seminar in the Theory and History of Media. I thank the organizers and participants for fruitful conversation and Reinhold Martin for his astute commentary.

For reading all or significant parts of the manuscript, I thank Stephanie Dick, Michael Harris, Stefan Helmreich, Matthew Jones, Colin McLarty, Christopher Phillips, Joan Richards, and Sophia Roosth. Stephanie Dick in particular has been thinking with me about mathematics, computation, and rationality from the time we were both students. Her comments on the complete draft of the manuscript were invaluable. Nick Liptak helped edit drafts of some of the chapters.

At the University of Chicago Press, Karen Darling has shepherded this project with kindness and patience.

Finally, I thank my family: my parents, Moshe Steingart and Ruth Jaffe, taught me to love books from an early age and have supported me in all my endeavors. I had not realized how much I was dependent on my visits home to see my sisters and all my nieces until the pandemic put a sharp stop to it all. Without their support, writing this book would have been impossible. For his love and friendship, I thank Offir Dagan. Mira Roosth came to play with the kids and be an all-around best aunt when I needed to write.

Lev and Sasha each did their best to ensure this book would never see the light of day. They taught me to appreciate any minute I got to write this book and cherish even more my time away from it. Having completed the work in the middle of a global pandemic, I am incredibly thankful for all the daycare teachers and staff who took care of Lev and Sasha while I was typing away at my computer (The Oxford Street Day Care Co-Op, The Weekday School, and Tompkins Hall Nursery and Childcare). Finally,

I have been thinking with Sophia Roosth for longer than I can remember. She has read every word in the book (often more than twice) and was kind enough to entertain me every time I announced I needed to start all over again. For sometimes bending space-time itself to support my work, for her love and kindness over the years, and for building a family with me, I will forever be grateful.

Archival Collections

AAAP Abraham Adrian Albert Papers. Special Collection Research Center, University of Chicago, Chicago, IL.

AMSR American Mathematical Society, *Records 1888*. Call Numbers Ms. 75.2, 75.5., 75.6, 75.7, 75.8 John Hay Library, Brown University Archives, Providence, RI.

GBP George Birkhoff Papers, 1902–1946.

JWTP John W. Tukey Papers, 1937–2000. Call Number Mss. Ms. Coll 117. American Philosophical Society Library, Philadelphia, PA.

MMP Papers of Harold Marston Morse, 1892–1977. Call Number HUGFP 106.10. Pusey Library, Harvard University Archives, Cambridge, MA.

OVP Oswald Veblen Papers, 1881–1960. MSS44016. Manuscript Division, Library of Congress, Washington, DC.

PMC Oral History Project archived website, "The Princeton Mathematics Community in the 1930s"; Department of Mathematics Oral History Project records, AC057, Princeton University Archives, Department of Special Collections, Princeton University Library, Princeton, NJ.

RAMP Records of the Applied Mathematics Panel, 1942–1946, Records of Panels and Committees, Records of the Office of Scientific Research and Development (OSRD). Call Number 227.5.4. National Archives at College Park, MD.

RCP Richard Courant Papers, 1902–1972. Call Number MC 150, Elmer Holmes Bobst Library, New York University Archives, New York. While I was visiting the archive the collection was being rearranged. Box X (new), refers to material that had already been processed according to the new organization. Box Y (old) refers to the older system.

RWP Raymond Wilder Papers, 1914–1982. Dolph Briscoe Center for American History, University of Texas at Austin, Austin, TX.

SMP Saunders Mac Lane Papers, 1969–1979. Dolph Briscoe Center for American History, University of Texas at Austin, Austin, TX.

SMR School of Mathematics Records, 1935–2016. Shelby White and Leon Levy Archives Center, Institute for Advanced Study, Princeton, NJ.

Notes

Introduction

1. Hermann Weyl, "Address at the Princeton Bicentennial Conference," in *Mind and Nature: Selected Writings on Philosophy, Mathematics, and Physics*, ed. Peter Pesic (Princeton, NJ: Princeton University Press, 2009), 170–71.

2. Gombrich wanted to name the article "The Vogue of Abstract Art," but the *Atlantic*'s editors insisted on the title. When Gombrich reprinted the essay in his book, he did so under his preferred title. Ernst Gombrich, "The Tyranny of Abstract Art," *Atlantic Monthly* 201 (1958): 43–48; Ernst Gombrich, *Meditations on a Hobby Horse: And Other Essays on the Theory of Art* (New York: Phaidon, 1963).

3. Ernst H. Gombrich, *The Story of Art* (New York, Phaidon, 1957), 445.

4. Gombrich, "The Vogue of Abstract Art," 45. On Gombrich's view of abstract art, see Caroline Jones, *Eyesight Alone: Clement Greenberg's Modernism and the Bureaucratization of the Senses* (Chicago: University of Chicago Press, 2005), 97–143.

5. A list of the participants in the conference serves as a who's who of American postwar social sciences. The attendees included Herbert Simon, David Riesman, Harold Lasswell, Philip M. Hauser, Walter A. Weisskopf, Talcott Parsons, Fred Eggan, George Stigler, Jacob Viner, Leonard White, Walter Johnson, Leo Strauss, and Samuel Stouffer, among many many more. See Leonard D. White, *The State of the Social Sciences* (Chicago: University of Chicago Press, 1956).

6. F. A. Hayek, "The Dilemma of Specialization," in *The State of the Social Sciences*, ed. Leonard D. White (Chicago: University of Chicago Press, 1956), 463.

7. Hayek had voiced these ideas already in the 1940s. See: F. A. Hayek, "Scientism and The Study of Society: Part I," *Economica* 9, no. 35 (1942): 267–91; F. A. Hayek, "Scientism and The Study of Society: Part II," *Economica* 10, no. 37 (1943): 34–63; F. A. Hayek, "Scientism and the Study of Society: Part III," *Economica* 11, no. 41 (1944): 27–39. He later republished these papers in *The Counter-Revolution of Science: Studies on the Abuse of Reason* (Indianapolis: Liberty Press, 1979).

8. Hayek, "Dilemma of Specialization," 463.

9. As Joel Isaac writes, theory in the social sciences began to be pursued in the 1930s, when a host of developments "did not induce social scientists to abandon their empiricist predilections, but they legitimized the idea that theoretical abstraction has a role to play in the discovery of social laws." Joel Isaac, "Tangled Loops: Theory, History, and the Human Sciences in Modern America," *Modern Intellectual History* 6, no. 2 (2009): 410.

10. Salomon Bochner, "Why Mathematics Grows," *Journal of the History of Ideas* 26, no. 1 (1965): 24.

11. Bochner, 24.

12. David Hilbert began advocating for the axiomatization of physics as early as 1900, when he added it to his twenty-three problems at the International Congress of Mathematics in Paris. Leo Corry, *David Hilbert and the Axiomatization of Physics (1898–1918): From* Grundlagen der Geometrie *to* Grundlagen der Physik (New York: Springer Science+Business Media, 2004).

13. Theodore M. Porter, *Trust in Numbers: The Pursuit of Objectivity in Science and Public Life* (Princeton, NJ: Princeton University Press, 1996).

14. In *The Making of the Cold War*, for example, Ron Robin argues that the mathematization of the human sciences filled a crucial need in the postwar period: "The Cold War was terra incognita and in need of a map, however tenuous; the application of a grid of mathematics fulfilled this cartographic lacuna." Such an explanation, however, runs the risk of treating mathematics as an empty signifier. It assumes that mathematics is merely a stable and timeless body of knowledge that can be called upon at will and put to any desired end. If we wish to understand what distinguishes a mathematical theory of human behavior in the nineteenth century from one in the twentieth century, it stands to reason that, at least in part, this difference would be due to changes in mathematics. Ron Theodore Robin, *The Making of the Cold War Enemy: Culture and Politics in the Military-Intellectual Complex* (Princeton, NJ: Princeton University Press, 2009), 71.

15. What follows is obviously not exhaustive, nor could it be. Rather it is a quick historical gloss intended to orient the reader.

16. Jonathan Lear, "Aristotle's Philosophy of Mathematics," *Philosophical Review* 91, no. 2 (1982): 161–92; John J. Cleary, "On the Terminology of 'Abstraction' in Aristotle," *Phronesis* 30, no. 1 (1985): 13–45; Edward Hussey, "Aristotle on Mathematical Objects," *Apeiron* 24, no. 4 (1991): 105–34.

17. Paul Vincent Spade, trans., *Five Texts on the Mediaeval Problem of Universals: Porphyry, Boethius, Abelard, Duns Scotus, Ockham* (Indianapolis: Hackett, 1994), 24.

18. Ignacio Angelelli, "Adventures of Abstraction," *Poznan Studies in the Philosophy of the Sciences and the Humanities* 82 (2004): 11–36; Ignacio Angelelli, "The Troubled History of Abstraction," ed. Uwe Meixner and Albert Newen, *Logical Analysis and History of Philosophy* 8 (2005): 157–75.

19. Margaret Atherton, "Berkeley's Anti-Abstractionism," in *Essays on the Philosophy of George Berkeley*, ed. Ernest Sosa (Boston: D. Riedel, 1987), 45–60; Douglas M. Jesseph, "Berkeley's Philosophy of Mathematics," in *The Cambridge Companion to Berkeley*, ed. Kenneth P. Winkler (Cambridge: Cambridge University Press, 2005), 266–310.

20. Atherton, "Berkeley's Anti-Abstractionism."

21. George Berkeley, *A Treatise Concerning the Principles of Human Knowledge*, ed. Kenneth Winkler (Indianapolis: Hackett, 1982), 9.

22. John Locke, *An Essay Concerning Human Understanding*, ed. Kenneth P. Winkler (Indianapolis: Hackett, 1996), book 4, chap. 7, para. 9, 267. Emphasis added.

23. Berkeley, *Principles of Human Knowledge*, 17.

24. Vere Chappell, "Locke's Theory of Ideas," in *The Cambridge Companion to Locke (Cambridge Companions to Philosophy)*, ed. Vere Chappell (Cambridge: Cambridge University Press, 1994), 26–55.

25. David Hume, *A Treatise of Human Nature*, vol. 1, ed. T. H. Green and T. H. Grose (London: Longmans Green and Co., 1882), part 1. sec. 7, 325.

26. Hume, vol. 1, part 1, sec. 7, 328.

27. Richard Bourke, *Empire and Revolution: The Political Life of Edmund Burke* (Princeton, NJ: Princeton University Press, 2015), 154.

28. As Bourke shows, Burke's suspicion of abstraction in political affairs predated the publication of *Reflections on the Revolution*. In his 1775 speech on conciliation with America, Burke attacked those who offered a speculative solution to the problem. According to Bourke, Burke held that "a new plan could not be based on unworldly abstraction, nor constructed on the basis of geometrical reasoning, but had to be grounded in evidence and fact." Bourke, 477.

29. Berkeley also rails against the philosophers and theologians who spent countless hours disputing the notions of abstract ideas. "What bickerings and controversies, and what a learned dust have been raised about those matters, and what mighty advantage has been from thence derived to mankind, are things at this day too clearly known to need being insisted on." Berkeley, *Principles of Human Knowledge*, 16.

30. Edmund Burke, *Reflections on the Revolution in France*, ed. L. G. Mitchell, Oxford World's Classics (New York: Oxford University Press, 1993), 61.

31. Burke, 60.

32. Burke, 60.

33. Burke, 39.

34. See, for example: Patrick Murray, *Marx's Theory of Scientific Knowledge* (Atlantic Highlands, NJ: Humanities Press International, 1988); Alberto Toscano, "The Open Secret of Real Abstraction," *Rethinking Marxism* 20, no. 2 (2008): 273–87; Patrick Murray, "Marx, Berkeley and Bad Abstractions," in *Marx and Contemporary Critical Theory: The Philosophy of Real Abstraction*, ed. Antonio Oliva,

Ángel Oliva, and Iván Novara (Switzerland: Palgrave Macmillan, 2020), 129–49; Alfonso Maurizio Iacono, "Marx's Method and the Use of Abstraction," in *Marx and Contemporary Critical Theory: The Philosophy of Real Abstraction*, ed. Antonio Oliva, Ángel Oliva, and Iván Novara (Switzerland: Palgrave Macmillan, 2020), 79–96; Paul Paolucci, "Marx's Method of Successive Abstractions and His Analysis of Modes of Production," *Critical Sociology* 46, no. 2 (2020): 171–89.

35. Karl Marx, *Grundrisse: Foundations of the Critique of Political Economy (Rough Draft)*, trans. Martin Nicolaus (London: Penguin, 1973), 104.

36. Marx, *Grundrisse*.

37. Fredric Jameson, *A Singular Modernity: Essay on the Ontology of the Present* (New York: Verso, 2002), 13.

38. Yve-Alain Bois et al., "Abstraction, 1910–1925: Eight Statements," *October* 143 (Winter 2013): 7.

39. Kazimir Malevich, *The Non-Objective World* (Chicago: Paul Theobald, 1959), 68.

40. Leah Dickerman, ed., *Inventing Abstraction, 1910–1925: How a Radical Idea Changed Modern Art* (New York: Museum of Modern Art, 2012), 14.

41. This, of course, does not imply that abstraction had a coherent meaning. The exhibit's catalogue, which includes essays from renowned scholars, makes clear that the meanings of abstraction were multiple. Abstraction might refer to, among other things, freedom and autonomy, the expression of one's inner workings, and the elementary constituents of painting, such as color and shape. It represented, as Hal Foster has argued, both "utopian-anarchic impulses" and "traditional-conservative commitments." Modernist abstraction, as the essays make clear, was a messy concept. Dickerman, *Inventing Abstraction, 1910–1925*.

42. Hubert Damisch and Stephen Bann, "Hubert Damisch and Stephen Bann: A Conversation," *Oxford Art Journal* 28, no. 2 (2005): 169.

43. Scholars across various fields have questioned the connection between transformations in science and in art at the turn of the century. Linda Dalrymple Henderson has shown how mathematicians' investigations into higher dimensions influenced artistic practice: Linda Dalrymple Henderson, *The Fourth Dimension and Non-Euclidean Geometry in Modern Art* (Cambridge, MA: MIT Press, 2013). John Adkins Richardson points to similarities between scientific thought and artistic vision, while Paul C. Vitz and Arnold B. Glimcher have focused on perception as a common concern for both scientists and artists: John Adkins Richardson, *Modern Art and Scientific Thought* (Chicago: University of Illinois Press, 1971); Paul C. Vitz and Arnold B. Glimcher, *Modern Art and Modern Science: The Parallel Analysis of Vision* (New York: Praeger, 1984). Robert Brain similarly points to perception, but focuses more directly on experimental physiology and instrumentation: Robert Michael Brain, *The Pulse of Modernism: Physiological Aesthetics in Fin-de-Siècle Europe* (Seattle: University of Washington Press, 2015). In *Inventing Abstraction*, scholars from various fields homed in on the changing meaning of abstraction to think through artistic and scientific practice at the turn of the century:

Dickerman, *Inventing Abstraction, 1910–1925*. More recently, Andrea Henderson has focused on Victorian society, demonstrating how mathematical formalism influenced literature, poetry, and photography: Andrea K. Henderson, *Algebraic Art: Mathematical Formalism and Victorian Culture* (Oxford: Oxford University Press, 2018). From the history of mathematics, see: Jeremy Gray, *Plato's Ghost: The Modernist Transformation in Mathematics* (Princeton, NJ: Princeton University Press, 2008).

44. Ferdinand de Saussure, *Course in General Linguistics*, ed. Charles Bally and Albert Sechehaye, trans. Wade Baskin (New York: Philosophical Library, 1959), 67.

45. Gerald Holton, "Einstein's Model for Constructing a Scientific Theory," in *Albert Einstein: His Influence on Physics, Philosophy and Politics*, ed. Peter C. Aichelburg and Roman U. Sexl (Braunschweig, Ger.: Friedr. Vieweg & Sohn, 1979), 109–36.

46. Paul Arthur Schilpp, ed., *Albert Einstein: Philosopher-Scientist* (Evanston, IL: Library of Living Philosophers, 1949), 89.

47. Alfred North Whitehead, *Science and the Modern World* (New York: Free Press, 1967 [1925]), 16.

48. Whitehead, 18.

49. Moritz Epple, "The End of the Science of Quantity: Foundations of Analysis, 1860–1910," in *A History of Analysis*, ed. Hans Niels Jahnke (Providence: American Mathematical Society, 2003), 291–324; Jesper Lützen, "The Foundation of Analysis in the 19th Century," in *A History of Analysis*, ed. Hans Niels Jahnke (Providence: American Mathematical Society, 2003), 155–96.

50. For an accessible history of numbers, see: Leo Corry, *A Brief History of Numbers* (Oxford: Oxford University Press, 2015).

51. In the German tradition, *Anschauung* denoted knowledge that was given to the senses that did not require any rational reconstruction in thought. According to Herbert Mehrtens, the meaning of *Anschauung* in mathematics was fairly stable and followed Kantian tradition. In his analysis of the Modernist transformation in mathematics, Mehrtens argues that modern mathematicians distinguished themselves from their predecessors by rejecting *Anschauung*. Herbert Mehrtens, *Moderne sprache mathematik: Eine geschichte des streit um die grundlagen der disziplin und des subjekts formaler system* (Frankfurt am Main: Suhrkamp, 1990).

52. For example, see Helmholtz's *Zählen und messen* (1887), Husserl's *Philosophie der arithmetik* (1891), Dedekind's *Was sind und was sollen die zahlen* (1888), Peano's *Arithmetices principia: nova methodo exposita* (1889), Frege's *Die grundlagen der arithmetik* (1884), and Russell's *The Principles of Mathematics* (1903). For English translations, see: Hermann Helmholtz, "Numbering and Measuring from an Epistemological Viewpoint," in *From Kant to Hilbert*, vol. 2, ed. William Bragg Ewald (Oxford: Oxford University Press, 2007), 727–52; Edmund Husserl, *Philosophy of Arithmetic: Psychological and Logical Investigations—with Supplementary Texts from 1887–1901*, trans. Dallas Willard (Berlin: Kluwer Academic Publishers, 2003); Richard Dedekind, *Essays on the Theory of Numbers*, trans. W. W. Beman, (New York: Dover, 1963); Giuseppe Peano, "The Principles of

Arithmetic, Presented by a New Method," in *From Frege to Gödel: A Source Book in Mathematical Logic, 1879–1931*, ed. Jean Van Heijenoort (Cambridge, MA: Harvard University Press, 1967), 83–97; Gottlob Frege, *The Foundations of Arithmetic: A Logico-Mathematical Enquiry into the Concept of Number*, trans. J. L. Austin (Evanston, IL: Northwestern University Press, 1980).

53. Hilbert was not alone in drawing attention to axiomatics. Italian mathematician Giuseppe Peano also made important contributions. See Ivor Grattan-Guinness, *The Search for Mathematical Roots 1870–1940: Logics, Set Theories and the Foundations of Mathematics from Cantor through Russell to Gödel* (Princeton: Princeton University Press, 2000).

54. On the debate between the two, see: Michael David Resnik, "The Frege-Hilbert Controversy," *Philosophy and Phenomenological Research* 34, no. 3 (1974): 386–403; Patricia A. Blanchette, "Frege and Hilbert on Consistency," *Journal of Philosophy* 93, no. 7 (July 1996): 317–36; Patricia Blanchette, "The Frege-Hilbert Controversy," *Stanford Encyclopedia of Philosophy*, ed. Edward N. Zalta, Spring 2014, http://plato.stanford.edu/archives/spr2014/entries/frege-hilbert/.

55. Gottlob Frege, *Philosophical and Mathematical Correspondence*, ed. Gabriel Gottfried et al. (Chicago: University of Chicago Press, 1980), 36.

56. Frege, *Philosophical and Mathematical Correspondence.*

57. Frege, 39.

58. Frege, 42. My emphasis.

59. Frege, 51.

60. When Oswald Veblen reviewed the work in *The Monist* in 1903, he noted, "Since its appearance in 1899 Hilbert's work on *The Foundations of Geometry* has had a wider circulation than any other modern essay in the realms of pure mathematics." Oswald Veblen, "Hilbert's Foundations of Geometry," *The Monist* 13, no. 2 (1903): 303.

61. Garrett Birkhoff and Mary Katherine Bennett, "Hilbert's 'Grundlagen Der Geometrie,'" *Rendiconti Del Circolo Matematico Di Palermo* 36, no. 3 (September 1987): 343.

62. On the difference between Hilbert's use of axiomatics and that of later mathematicians, see Leo Corry, "Axiomatics between Hilbert and the New Math: Diverging Views on Mathematical Research and Their Consequences on Education," *International Journal for the History of Mathematics Education* 2, no. 2 (2007): 21–37.

63. On Moore and the growth of the American mathematical community, see Karen Hunger Parshall, "Eliakim Hastings Moore and the Founding of a Mathematical Community in America, 1892–1902," *Annals of Science* 41 (1984): 313–33; Karen Hunger Parshall and David E. Rowe, *The Emergence of the American Mathematical Research Community, 1876–1900* (Providence: American Mathematical Society, 1997).

64. Eliakim Hastings Moore, "On the Foundations of Mathematics," *Science* 17, no. 428 (1903): 401.

65. David Hilbert, "Axiomatic Thought," in *From Kant to Hilbert: A Source Book in the Foundations of Mathematics*, vol. 2, ed. William Ewald (Oxford: Clarendon Press, 1996), 1105–15.

66. Mehrtens, *Moderne sprache mathematik*; Jeremy Gray, *Plato's Ghost: The Modernist Transformation of Mathematics* (Princeton, NJ: Princeton University Press, 2008); I. Grattan-Guinness, "Mathematics Ho! Which Modern Mathematics Was Modernist?" *Mathematical Intelligencer* 31, no. 4 (2009): 3–11.

67. Nicholas Bourbaki, "The Architecture of Mathematics," *American Mathematical Monthly* 57, no. 4 (1950): 228.

68. Leo Corry, *Modern Algebra and the Rise of Mathematical Structures* (Boston: Springer, 2004).

69. When Marshall Stone was given the reins of the Department of Mathematics at the University of Chicago, his first task was to rebuild its faculty. This was seen as a crucial step in the transformation of the department into a leading research center. One of his first hires was André Weil, one of Bourbaki's founders. Weil was not the only member to take a position in an American university. In 1945, Claude Chevalley, another of the group's five founders, took a position at Princeton University, and then two years later moved to Columbia University, where he stayed for three years before moving back to Paris. During the 1950s, several other members of Bourbaki's second and third generations also relocated to top American universities. The influence of Bourbaki on the development of American mathematics in the first two decades following World War II was not simply a matter of American mathematicians absorbing its philosophical views. The influence trafficked both ways. In 1957 and 1958, when the Institute for Advanced Study (IAS) hired two Bourbaki mathematicians, Weil and Armand Borel, the move solidified Bourbaki's influence while also confirming its status and importance. Stated somewhat differently, the impact of Bourbaki on the US was not inevitable; it had to be produced and maintained.

70. James C. Scott, *Seeing Like a State: How Certain Schemes to Improve the Human Condition Have Failed* (New Haven, CT: Yale University Press, 1998), 90.

71. Hunter Heyck, "The Organizational Revolution and the Human Sciences," *Isis* 105, no. 1 (March 2014): 9.

72. In the arts it is more difficult to get a handle on the term "high modernism," as scholars have used it to denote different movements. In literature, high modernism primarily describes literature of the interwar period, exemplified by the works of T. S. Eliot, Robert Musil, Virginia Woolf, and Thomas Mann. However, the term "late modernism" has also been used to describe some of the same works (see Tyrus Miller). Expanding beyond the strict confines of literature, Robert Genter has suggested high modernism as a way of describing the work of literary critics, art critics, and New York intellectuals writing during the first decades after World War II. According to Genter, despite their diverse backgrounds, these intellectuals were united in their opposition to the encroachment of scientific reasoning into every domain of social life. In elevating humanistic studies, they were further united in maintaining the

autonomy of the aesthetic. As such, Genter's high modernists correspond strongly with those Fredric Jameson denoted as late modernists (as he distinguished them from high modernists). Joshua Kavaloski, *High Modernism: Aestheticism and Performativity in Literature of the 1920s* (Rochester, NY: Boydell & Brewer, 2014); Robert Genter, *Late Modernism: Art, Culture, and Politics in Cold War America* (Philadelphia: University of Pennsylvania Press, 2010); Jameson, *A Singular Modernity*.

73. Genter, *Late Modernism*, 37.

74. Jameson, *A Singular Modernity*, 176.

75. David A. Hollinger, "The Canon and Its Keepers: Modernism and Mid-Twentieth-Century American Intellectuals," in *In the American Province: Studies in the History and Historiography of Ideas* (Baltimore: Johns Hopkins University Press, 1989), 74–91.

76. The relation between positivism and abstraction goes all the way back to French philosopher Auguste Comte's classification of the sciences. For Comte, the sciences were arranged in succession according to levels of generality and simplicity. Thus, psychology is dependent on biology, which in turn is dependent on chemistry, and hence upon physics. Biological phenomena were more complex than those of chemistry, but were not as generalized. Reigning above all was mathematics. "In placing mathematics at the head of the Positive Philosophy," Comte explained, "we are only extending the application of the principle which had governed our whole Classification." Namely, mathematical phenomena were "the most general, the most simple, the most abstract of all,—the most irreducible to others, the most independent of them." Comte is usually credited as the "father" of sociology, having included social physics within his classification of the sciences (it appears on the bottom, as the least generalized but most complex). In his *Cours de philosophie positive*, he claimed that every phenomenon can be reduced to number, and is hence open to analytical investigation. It took almost a century for Comte's prescription for the social sciences to come to fruition. Comte insisted that it was through the application of mathematics to social phenomena that the science would be elevated into its positive stage. Yet, by the time this transformation finally took place, what was involved in such mathematization had itself changed. It was not quantification per se, the reduction of the phenomena to number à la Comte, but rather the formalization (and axiomatization) that governed the process. Mathematics changed, as did the meaning of abstraction. Auguste Comte, *The Positive Philosophy*, trans. Harriet Martineau (New York: Calvin Blanchard, 1858), 50.

77. For two wonderful review essays, see Joel Isaac, "The Human Sciences in Cold War America," *Historical Journal* 50, no. 3 (2007): 725–46; Nils Gilman, "The Cold War as Intellectual Force Field," *Modern Intellectual History* 13, no. 2 (2016): 507–23.

78. This is also why my story ends in the mid-1970s, when postwar expansion came to an end and the entire rationale for constructing it was called into question.

79. Another crucial question has to do with the periodization of the Cold War. Anders Stephanson has forcefully argued that the Cold War has become an empty

signifier, a term that has been used to describe any foreign policy activity that occurred between the end of World War II and 1989. I limit my use of the term "Cold War" throughout the book to those times when I want the concept to highlight the particular conditions of knowledge at the time that were derived (even to a second or third degree) from the political agenda of the Cold War. Anders Stephanson, "Fourteen Notes on the Very Concept of The Cold War," in *Rethinking Geopolitics*, ed. Simon Dalby and Gearóid Ó. Tuathail (New York: Routledge, 1998), 62–85; Anders Stephanson, "Cold War Degree Zero," in *Uncertain Empire: American History and the Idea of the Cold War*, ed. Joel Isaac and Duncan Bell (New York: Oxford University Press, 2012), 19–50.

Chapter One

1. In the end, neither of the two mathematicians joined Harvard. Chevalley chose to stay at Columbia, and Beurling immigrated to the US in 1954 and was appointed as a professor at the Institute for Advanced Study at Princeton.

2. On Bourbaki, see David Aubin, "The Withering Immortality of Nicolas Bourbaki: A Cultural Connector at the Confluence of Mathematics, Structuralism, and the Oulipo in France," *Science in Context* 10, no. 2 (1997): 297–342; Liliane Beaulieu, "Dispelling a Myth: Questions and Answers about Bourbaki's Early Work, 1934–1944," in *The Intersection of History and Mathematics*, ed. Sasaki Chikara, Sugiura Mitsuo, and Joseph W. Dauben (Boston: Birkhäuser-Verlag, 1994), 241–52; Liliane Beaulieu, "Bourbaki's Art of Memory," *Osiris* 14 (1999): 219–51; Leo Corry, "Nicolas Bourbaki and the Concept of Mathematical Structure," *Synthese* 92, no. 3 (1992): 315–48. For recollections by first generation Bourbaki, see Armand Borel, "Twenty-Five Years with Nicolas Bourbaki, 1940–1973," *Notices of the American Mathematical Society* 45, no. 3 (1998): 373–80; Jean A. Dieudonné, "The Work of Nicholas Bourbaki," *American Mathematical Monthly* 77, no. 2 (1970): 134–45.

3. Zariski, Oscar, "A Statement in Support of Claude Chevalley Candidacy for a Permanent Appointment." OVP, box 23, folder "Harvard 1951." Garrett Birkhoff, another Harvard member, similarly praised Chevalley for his "originality," noting that he had "a real feeling for abstract essence," and that his "interests [were] wide." Garrett Birkhoff to Joe Walsh, December 20, 1950. OVP, box 23, folder "Harvard 1951."

4. Ahlfors, Lars. "Comment on the Department's Recommendation for a Permanent Appointment." OVP, box 23, folder "Harvard 1951."

5. This is not to say that intuition and logic stopped being a common refrain for mathematicians. The point is, rather, that the opposition between abstraction and concreteness took precedent.

6. Fredric Jameson, *A Singular Modernity: Essay on the Ontology of the Present* (New York: Verso, 2002), 40.

7. Jameson, *A Singular Modernity*, 35.

8. Jameson, 36.

9. Richard Courant, "Mathematics in the Modern World," *Scientific American* 211, no. 3 (1964): 44.

10. David E. Zitarelli, "The Origin and Early Impact of the Moore Method," *American Mathematical Monthly* 111, no. 6 (June, 2004): 465–86; John Parker, *R.L. Moore: Mathematician and Teacher* (Washington, DC: Mathematical Association of America, 2005).

11. Albert C. Lewis, "The Beginnings of the R.L. Moore School of Topology," *Historia Mathematica* 31, no. 3 (August 2004): 279–95.

12. Topology can roughly be defined as studying the invariants of space under continuous transformations. The common joke that, for a topologist, a doughnut and a coffee mug are indistinguishable centers around the fact that the doughnut "hole" is exactly such an invariant. No matter how you continuously deform the surface of the doughnut, if you neither break nor add anything to it, it will always have exactly one hole—like the single hole inside the handle of the coffee cup. You can neither get rid of it nor add an additional one.

13. On the distinction between point-set and algebraic topology, see I. M. James, "Combinatorial Topology versus Point-Set Topology," in *Handbook of the History of General Topology*, ed. C. E. Aull and R. Lowen (Dordrecht, Neth.: Kluwer Academic Publishers, 2001), 809–34. For a technical history of topology, see: I. M. James, *History of Topology* (Amsterdam: Elsevier, 1999); Jean Dieudonné, *A History of Algebraic and Differential Topology, 1900—1960* (Boston: Springer Science+Business Media, 2009).

14. The correspondence, which spans approximately a decade, consists of dozens of letters. The majority of the letters cover the period from 1932 to 1936 and include letters from both Wilder and Steenrod. It is during these early years that Steenrod matured in mathematics, and throughout the period he kept Wilder informed about his progress. The entire correspondence can be found in RWP, 3.6/86–36/13.

15. R. L. Wilder to President A. Lawrence Lowell, February 25, 1933.

16. *Acyclical* curves were continuous curves that did not contain any simple closed curve. Norman Steenrod to Raymond Wilder, June 15, 1933.

17. Norman Steenrod to Raymond Wilder, October 19, 1933.

18. Steenrod, to Wilder, October 19, 1933.

19. Steenrod to Wilder, December 10, 1933.

20. In Veblen's obituary, Mac Lane notes that "for many years [*Analysis Situs*], as the best source, was assiduously studied by generations of topologists who have gradually wholly transformed the subject." Saunders Mac Lane, "Oswald Veblen," *Biographical Memoirs, National Academy of Sciences* 37 (1964): 329. Hassler Whitney similarly remarked that Veblen's book "had an enormous influence in increasing interest in the subject, and was the standard text for a decade." Hassler Whitney, "Review: Solomon Lefschetz, *Algebraic Topology*," *Mathematical Reviews* (1943), MR0007093.

21. Steenrod to Wilder, February 16, 1934.

22. Steenrod to Wilder, March 31, 1934.

23. Steenrod to Wilder, June 9, 1934.

24. Steenrod to Wilder, June 9, 1934.

25. Wilder to Steenrod, September 4, 1934.

26. Among those mathematicians who were at Princeton during the years Steenrod was there were Albert Tucker, Nathan Jacobson, Robert Walker, Merrill Flood, Abraham Taub, and Carl Allendoerfer.

27. Saunders Mac Lane, "Topology and Logic at Princeton," in *A Century of Mathematics in America*, part 2, ed. Peter L. Duren (Providence: American Mathematical Society, 1988), 220.

28. Emphasis added. Steenrod to Wilder, November 18, 1934.

29. Both the Polish school and Moore's school focused on set theoretical topology, although they differed in the type of problems and approach they advocated.

30. Steenrod to Wilder, November 18, 1934.

31. Steenrod to Wilder, November 18, 1934.

32. William L. Duren, Nathan Jacobson, and Edward J. McShane. PMC, transcript number 8.

33. This distinction is very much alive today. In 2000, mathematician Timothy Gowers published *The Two Cultures of Mathematics* in which he adopts C. P. Snow's two culture argument from 1959 to describe the current mathematical community. "The 'two cultures' I wish to discuss will be familiar to all professional mathematicians. Loosely speaking, I mean the distinction between mathematicians who regard their central aim as being to solve problems, and those who are more concerned with building and understanding theories." William Timothy Gowers, "The Two Cultures of Mathematics," in *Mathematics: Frontiers and Perspectives*, ed. V. Arnold et al. (Providence: American Mathematical Society, 2000), 65–78.

34. Among Lefschetz's students (many of whom feature throughout this book) were John Tukey, Albert Tucker, Clifford Truesdell III, Arthur Stone, Paul Smith, Ralph Fox, William Flexner, Felix Browder, Richard Bellman, and Edward Begle.

35. The best source on the transformation in algebra is Leo Corry, *Modern Algebra and the Rise of Mathematical Structures* (Boston: Springer, 2004).

36. Steenrod to Wilder, November 18, 1934.

37. Steenrod to Wilder, November 18, 1934.

38. Évariste Galois first used the concept of a group in 1830, but its meaning did not stabilize until the last decade of the century. As Hans Wussing has demonstrated, the concept of an abstract group arose slowly and was the cumulative work of many mathematicians. Intuitively, one can think of a group as a way of defining a symmetrical relationship. For example, the *symmetric group* designates the various permutations and combinations of permutations one can perform on a set of *n* elements. See: Hans Wussing, *The Genesis of the Abstract Group Concept: A Contribution to the History of the Origin of Abstract Group Theory* (Cambridge, MA: MIT Press, 1984).

39. Corry, *Modern Algebra*, 248.

40. Anna Pell Wheeler to Oswald Veblen, August 7, 1933, OVP, box 9, folder "Noether."

41. William L. Duren, Nathan Jacobson, and Edward J. McShane. PMC, transcript number 8.

42. Garrett Birkhoff, "Current Trends in Algebra," *American Mathematical Monthly* 80, no. 7 (1973): 771.

43. In "The Suppressed Drawing: Paul Dirac's Hidden Geometry," Peter Galison writes about the stark difference between physicist Paul Dirac's private notes, which were filled with geometric drawings, and his published work, which was known "for the austerity of his prose, his rigorous fundamentally algebraic solution to every physical problem he approached." Galison reads the suppression of Dirac's drawings as a historical process by asking, "What are the specific conditions that govern the separation of certain practices from the public domain?" With Steenrod, the dichotomy between geometry and algebra does not mark a line between the public and the private. Rather, it reflects the development of topology itself. Steenrod did not continue to draw illustrations because the topological questions that occupied him by the time he received his doctorate simply did not lend themselves to any clear visualization. His entire approach and understanding of the field had shifted. This is what distinguished Steenrod's generation from the post–WWII one. As I demonstrate in this chapter, whereas the former was introduced to topology by first learning its geometric basis and only then shifting to the algebraic point of view, the latter approached topology as a fully algebraic field from the start. Thus, Steenrod's professional trajectory marks a broader transformation in the growth of the field. Peter Galison, "The Suppressed Drawing: Paul Dirac's Hidden Geometry," *Representations*, no. 72 (2000): 145.

44. Garrett Birkhoff, "Some Leaders in American Mathematics: 1891–1941," in *The Bicentennial Tribute to American Mathematics, 1776–1976*, ed. Dalton Tarwater (Washington, DC: Mathematical Association of America, 1976), 68.

45. Birkhoff, "Some Leaders," 68–69.

46. Lefschetz's book was not the only one to present the algebraic side of topology. Here, I am focusing only on the English language textbooks, but others were published as well. For example, P. Alexandroff and H. Hopf published *Topologie I*, and H. Seifert and W. Threlfall published *Lehrbuch der Topologie*.

47. George D. Birkhoff, "Review: Topology," *Science*, New Series 96, no. 2504 (December 25, 1942): 581–84.

48. Birkhoff, "Review: Topology."

49. Alma Steingart, "Conditional Inequalities: American Pure and Applied Mathematics, 1940–1975," (PhD diss., Massachusetts Institute of Technology, 2013), chap. 1.

50. Birkhoff, "Review: Topology," 582.

51. Note that the opposite is not true. If two spaces have the same algebraic concepts attached to them, it is not guaranteed that they can be continuously deformed to one another. That is, the condition is necessary but not sufficient.

52. In the seventh chapter of the book on homology theory, Lefschetz begins by noting that there are several known theories. For compact spaces there are "theories due to Vietoris, Alexanderoff and Lefschetz," while for general spaces there was a "theory due to Kurosch, and another due to (or rather patterned after one due to) Alexander and Kolmogoroff." He then added, "A complete discussion will also be given of the Kurosch and Alexander-Kolmogoroff types and they will be proved equivalent to the I-type. Thus at last a certain *unity* will have been brought into this group of questions." Solomon Lefschetz, *Algebraic Topology* (Providence: American Mathematical Society, 1942), 244.

53. Hassler Whitney, "Review: Solomon Lefschetz."

54. My emphasis. Whitney, "Review: Solomon Lefschetz."

55. Birkhoff, "Review: Topology," 582.

56. Birkhoff, 582.

57. Birkhoff, "Review: Topology."

58. Eilenberg was one of the many European mathematicians who immigrated to the United States in the 1930s out of fear of persecution. See Reinhard Siegmund-Schultze, *Mathematicians Fleeing from Nazi Germany: Individual Fates and Global Impact* (Princeton, NJ: Princeton University Press, 2009).

59. Samuel Eilenberg to Raymond Wilder, October 19, 1940, RWP, box 3.6/86–36/4.

60. Samuel Eilenberg and Norman E. Steenrod, "Axiomatic Approach to Homology Theory," *Proceedings of the National Academy of Sciences of the United States of America* 31, no. 4 (1945): 117–20.

61. Eilenberg and Steenrod, "Axiomatic Approach," 117.

62. Eilenberg and Steenrod, 117.

63. Samuel Eilenberg and Norman Steenrod, *Foundations of Algebraic Topology* (Princeton, NJ: Princeton University Press, 1952), viii.

64. See the introduction to the focus issue of *Isis* and the articles in it: Marga Vicedo, "Introduction: The Secret Lives of Textbooks," *Isis* 103, no. 1 (March 2012): 83–87. David Kaiser, ed., *Pedagogy and the Practice of Science: Historical and Contemporary Perspectives* (Cambridge, MA: MIT Press, 2005).

65. The first chapter of Lefschetz's book presented the basic concepts and theorems of geometric topology.

66. Eilenberg and Steenrod, *Foundations of Algebraic Topology*, viii.

67. For a technical development of homological algebra, see Charles Weibel, "History of Homological Algebra," in *History of Topology*, ed. I. M. James (New York: Elsevier, 1999), 797–836.

68. Henri Cartan and Samuel Eilenberg, *Homological Algebra* (Princeton, NJ: Princeton University Press, 1956), v.

69. Cartan and Eilenberg, *Homological Algebra*, v.

70. Saunders Mac Lane, "Review: Homological Algebra," *Bulletin of the American Mathematical Society* 62, no. 6 (November 1956): 622.

71. Mac Lane, 622.

72. MacLane, 622.

73. In his monumental book *The Logical Structure of the World: Pseudoproblems in Philosophy*, Rudolf Carnap announced, "Only if we succeed in producing such a unified system of all concepts will it be possible to overcome the separation of unified science to unrelated special sciences." Rudolf Carnap, *The Logical Structure of the World: Pseudoproblems in Philosophy*, trans. Rolf A. George (Los Angeles: University of California Press, 1967), 7.

74. Peter Galison, "The Americanization of Unity," *Daedalus* 127, no. 1 (1998): 67.

75. As such, their view was similar to one espoused by the prewar Unity of Science movement. Indeed, both movements emphasized axiomatics and logic in their constructions, with roots going back to foundational concerns at the turn of the century. However, there was a fundamental difference between the conception of unity for the prewar Unity of Science movement and mathematicians. In mathematics, unification did not serve as a philosophical imperative, but a practical one. The linguistic theories promoted by Carnap and Otto Neurath served as post hoc reconstructions of science. They provided a philosophical framework for many scientists, but they did not directly impact the constitution of their subject matter. This was not the case in mathematics, where unification functioned on the ground. It impacted the sorts of questions mathematicians considered, the approach they advocated, and the textbooks they wrote, until it eventually completely transformed what mathematics *was* as a field of study.

76. Saunders Mac Lane, "Concepts and Categories in Perspective," in *A Century of Mathematics in America*, ed. Peter Duren (Providence: American Mathematical Society, n.d.), 334. For a history of the development of category theory, see Corry, *Modern Algebra*, chap. 8.

77. Saunders Mac Lane, *Saunders Mac Lane: A Mathematical Autobiography* (Wellesley, MA: A K Peters, 2005), 99–104.

78. MacLane, *Saunders Mac Lane*, 125.

79. Hyman Bass et al., "Samuel Eilenberg (1913–1998)," *Notices of the American Mathematical Society* 45, no. 10 (November 1998): 1350.

80. Saunders Mac Lane, "The PNAS Way Back Then," *Proceedings of the National Academy of Sciences* 94 (June 1997): 5983.

81. Some mathematicians have also turned to category theory in order to provide a foundation for all of mathematics. See Colin McLarty, *Elementary Categories, Elementary Toposes* (Oxford: Clarendon Press, 1992); Colin McLarty, "Recent Debate over Categorical Foundations," in *Foundational Theories of Classical and Constructive Mathematics*, ed. Giovanni Sommaruga, Western Ontario Series in Philosophy of Science (Dordrecht: Springer Netherlands, 2011), 145–54.

82. Corry, *Modern Algebra*, 339.

83. Mac Lane, "Concepts and Categories," 335.

84. Samuel Eilenberg and Saunders MacLane, "General Theory of Natural Equivalences," *Transactions of the American Mathematical Society* 58, no. 2 (1945): 236.

85. Oscar Zariski, "A Statement in Support of Claude Chevalley Candidacy for a Permanent Appointment," OVP, box 23, folder "Harvard 1951."

86. This is, of course, a generalization. There were always mathematicians, even at the height of high modernist mathematics, who preferred, and dedicated their study to, concrete mathematical problems and were less interested in theorizing. However, on the whole, these mathematicians were the exception rather than the rule.

87. Raymond L. Wilder, "Heredity Stress as a Cultural Force in Mathematics," *Historia Mathematica* 1 (1974): 39.

88. In a footnote, Wilder comments that this situation involves "an interesting psychological problem": determining why older mathematicians tended to leave the field "when it reached higher levels of abstraction." Their motives, he suggests, could be that they were unable to "cope with the new," that they "just lost interest," or that they "seized the opportunity offered . . . to explore other fields." Most likely all of these reasons contributed to the phenomenon. Wilder, "Heredity Stress," 39.

89. Grothendieck is also one of the most intriguing figures of twentieth-century mathematics. In the 1960s he revolutionized the study of algebraic geometry, gaining many followers in France and abroad. He was one of the first professors at IHÉS (Institut des Hautes-Études Scientifiques), but then in the 1970s he abruptly resigned from his position and retreated from mathematical life. He became increasingly reclusive, cutting off all of his relationships from his earlier life. He has since become a source of fascination for many in the mathematical community. Winfried Scharlau, "Who Is Alexander Grothendieck?," *Notices of the AMS* 55, no. 8 (2008): 930–41; Leila Schneps, ed., *Alexander Grothendieck: A Mathematical Portrait* (Somerville, MA: International Press of Boston, 2014).

90. Colin McLarty, "The Rising Sea: Grothendieck on Simplicity and Generality," in *Episodes in the History of Modern Algebra (1800–1950)*, ed. Jeremy J. Gray and Karen Hunger Parshall (Providence: American Mathematical Society, 2011), 301–22.

91. Marston Morse, "Twentieth Century Mathematics," *American Scholar* 9, no. 4 (1940): 499–504.

92. Marston Morse to Arnaud Denjoy, January 14, 1971, MMP, box 4, folder "Denjoy."

93. Years later, Saunders Mac Lane made a similar point, noting that his own inclination toward generalization and axiomatization had been guided by the belief that new concepts are only effective if they can be applied in various contexts. "Parts of the mathematical literature," he noted, "are littered with such failed abstractions." Mac Lane, "Concepts and Categories," 329.

94. Raoul Bott, "Marston Morse and His Mathematical Works," *Bulletin of the American Mathematical Society* 3, no. 3 (1980): 907.

95. Bott, "Marston Morse."

96. Louis Joel Mordell, "Review: Serge Lang, Diophantine Geometry," *Bulletin of the American Mathematical Society* 70, no. 4 (1964): 495.

97. Mordell, "Review: Serge Lang," 497.

98. Serge Lang, "Mordell's Review, Siegel's Letter to Mordell, Diophantine Geometry, and 20th Century Mathematics," *Notices of the American Mathematical Society* 42, no. 3 (1995): 339.

99. "Carl Ludwig Siegel to Robert Oppenheimer," December 12, 1959, OVP, box 12, folder "Siegel."

100. "Carl Ludwig Siegel to Borel et al.," February 15, 1960, MMP, box 4, folder "Siegel."

101. "Carl Ludwig Siegel to Borel."

102. "Minutes of a Meeting of the Professor of the School of Mathematics Held February 2, 1945," SMR, faculty minutes, box 1, folder 1945–1953, 11.

103. "Minutes of a Meeting."

104. "Minutes of a Meeting," 12.

105. "Minutes of a Meeting," 9.

106. Washington Irving, *Rip Van Winkle and the Legend of Sleepy Hollow* (New York: Macmillan and Co., 1893), 84.

Chapter Two

1. J. L. Synge to Churchill Eisenhart et al., September 3, 1946, AMSR, box 31, folder 40.

2. Synge to Eisenhart, September 3, 1946.

3. John H. Curtiss to John Synge, November 6, 1946, AMSR, box 31, folder 41.

4. Curtiss to Synge, November 6, 1946.

5. The respondents varied in their suggestions. Nine mathematicians strongly objected to the establishment of a separate division for applied mathematics, eight supported the proposal, and three held that a completely new organization should be established. All the original members of the committee (except for von Neumann) voted against the establishment of a new division, although he changed his mind later and voted with the rest of the committee. In explaining their respective positions, mathematicians cited both intellectual and social reasons. For example, some pointed out that it was conceptually impossible to clearly distinguish between applied and pure mathematics, or that the different branches of applied mathematics had more in common with pure mathematics than with each other, and hence a separate division was unnecessary. Others noted that the assumed hierarchy between pure and applied mathematics, to which most pure mathemati-

cians subscribed, required a separate division if applied mathematics was to be developed properly.

6. Paul Erickson et al., *How Reason Almost Lost Its Mind: The Strange Career of Cold War Rationality* (Chicago: University of Chicago Press, 2013), 3–4.

7. Erickson et al., *How Reason Almost Lost*, 10.

8. Erickson et al., 44.

9. Marshall Stone to Marston Morse, August 6, 1941, MMP, box 13, folder "Stone."

10. Griffith Evans, the chair of the Department of Mathematics at the University of California, Berkeley, also helped prepare the memorandum. However, he did not travel to Washington for the meeting.

11. Marshall Stone, "Crisis at Harvard," *Nation*, December 2, 1939.

12. Marshall Stone and Marston Morse, "Mathematics in War," undated, MMP, box 13, folder "Stone."

13. Stone and Morse, "Mathematics in War."

14. Frank Jewett to Marston Morse, March 30, 1942, MMP, box 9, folder "Jewett."

15. Jewett to Morse, March 30, 1942.

16. On the disagreement that erupted, see Larry Owens, "Mathematics at War: Warren Weaver and the Applied Mathematics Panel 1942–1945," in *The History of Modern Mathematics: Institutions and Applications*, ed. David E. Rowe and John McCleary (Boston: Academic Press, 1988), 286–305; Alma Steingart, "Conditional Inequalities: American Pure and Applied Mathematics, 1940–1975" (PhD diss., Massachusetts Institute of Technology, 2013), chap. 1.

17. As Silvan Schweber has demonstrated, at the beginning of the twentieth century American physics took a practical outlook. Unlike their European counterparts, American physicists were involved in experimental work. In contrast, American mathematicians followed the German model more closely. Silvan S. Schweber, "The Empiricist Temper Regnant: Theoretical Physics in the United States 1920–1950," *Historical Studies in the Physical and Biological Sciences* 17, no. 1 (January 1986): 58.

18. Stone and Morse, "Mathematics in War."

19. Stone and Morse, "Mathematics in War." Emphasis in original.

20. Marshall Stone to Dunham Jackson, July 21, 1940, MMP, box 13, folder "Stone."

21. Stone to Jackson, July 21, 1940.

22. Stone to Jackson, July 21, 1940.

23. Marston Morse, "George David Birkhoff and His Mathematical Work," *Bulletin of the American Mathematical Society* 52, no. 5 (1946): 358.

24. The two elements Veblen used were a point and an order. Here he was following the work of German mathematician Moritz Pasch and Italian mathematician Giuseppe Peano. Oswald Veblen, "A System of Axioms for Geometry," *Transactions of the American Mathematical Society* 5, no. 3 (1904): 343–84.

25. The American interest in Hilbert's axiomatic analysis is often referred to as postulational analysis. For more on this work, see Michael Scanlan, "Who Were the American Postulate Theorists?" *Journal of Symbolic Logic* 56, no. 3 (1991): 981–1002.

26. On Veblen's influence, see William Aspray, "Oswald Veblen and the Origins of Mathematical Logic at Princeton," in Thomas Drucker, ed., *Perspectives on the History of Mathematical Logic* (London: Springer, 1991), 54–70; Loren Butler Feffer, "Oswald Veblen and the Capitalization of American Mathematics: Raising Money for Research, 1923–1928," *Isis* 89, no. 3 (1998): 474–97.

27. Oswald Veblen and John Wesley Young, *Projective Geometry* (Boston: Ginn and Company, 1910), 7.

28. Veblen and Young, *Projective Geometry*, 15.

29. On Veblen's efforts, see William Aspray, "Oswald Veblen," 54–70; Loren Butler Feffer, "Oswald Veblen and the Capitalization of American Mathematics: Raising Money for Research, 1923–1928, " *Isis* 89, no. 3 (1998): 474–97.

30. James Wallace Givens, Abraham H. Taub and Angus E. Taylor, PMC, transcript number 14.

31. George Birkhoff to Oswald Veblen, January 12, 1925, OVP, box 2, folder "Birkhoff."

32. Birkhoff to Veblen, January 12, 1925.

33. In the postwar period, the same set of questions would haunt social scientists.

34. Oswald Veblen, "Remarks on the Foundations of Geometry," *Bulletin of the American Mathematical Society* 31, no. 3–4 (1925): 135.

35. George David Birkhoff, *Relativity and Modern Physics* (Cambridge, MA: Harvard University Press, 1927).

36. George David Birkhoff, *The Origin, Nature, and Influence of Relativity* (New York: Macmillan, 1925), 176.

37. Birkhoff, *Origin, Nature, and Influence*, 176–77.

38. Quoted in Jimena Canales, *The Physicist and the Philosopher: Einstein, Bergson, and the Debate that Changed Our Understanding of Time* (Princeton, NJ: Princeton University Press, 2015), 19.

39. Oswald Veblen, "Geometry and Physics," *Science* 57, no. 1466 (1923): 131.

40. In his first review of relativity, Birkhoff approvingly evaluated Whitehead's *The Concept of Nature*. He then returned to Whitehead's writings in his own book on relativity, and while he approached the theory from a different perspective it is obvious that Birkhoff was influenced by Whitehead's philosophical ideas.

41. Here the language of "undefined" terms most clearly reveals the influence of Hilbert's axiomatics.

42. Birkhoff, *Origin, Nature, and Influence*, 178.

43. Veblen, "Remarks on the Foundations," 135.

44. Veblen, "Geometry and Physics," 131.

45. Oswald Veblen, "Certain Aspects of Modern Geometry," *Rice Institute Pamphlet* 21 (1934): 217.

46. George D. Birkhoff, "Intuition, Reason and Faith in Science," *Science* 88, no. 2296 (December 1938): 602.

47. Birkhoff, "Intuition, Reason and Faith," 602.

48. Birkhoff, "Intuition, Reason and Faith," 606.

49. George David Birkhoff, *Aesthetic Measure* (Cambridge, MA: Harvard University Press, 1933).

50. George David Birkhoff, "A Mathematical Approach to Ethics," *Rice Institute Pamphlet–Rice University Studies* 28, no. 1 (1941): 1–23.

51. George David Birkhoff, "Talk on the Present State of Scientific Talk," April 22, 1937, GBP HUG4213.4.

52. Leo Corry, "The Influence of David Hilbert and Hermann Minkowski on Einstein's Views over the Interrelation between Physics and Mathematics," *Endeavour* 22, no. 3 (1998): 95–97.

53. Quoted in Corry, "Influence of David Hilbert," 96.

54. Quoted in Gerald Holton, "Einstein's Model for Constructing a Scientific Theory," in *Albert Einstein: His Influence on Physics, Philosophy and Politics*, ed. Peter C. Aichelburg and Roman U. Sexl (Braunschweig, Ger.: Friedr. Vieweg & Sohn, 1979), 115.

55. Ronald N. Giere and Alan W. Richardson, ed., *Origins of Logical Empiricism* (Minneapolis: University of Minnesota Press, 1996), 8.

56. Carnap does bring the topic up, but only briefly.

57. Rudolf Carnap, *Foundations of Logic and Mathematics* (Chicago: University of Chicago Press, 1939), 2.

58. Carnap, *Foundations of Logic*, 64.

59. Carnap, *Foundations of Logic*, 67.

60. "The calculus is first constructed floating in the air, so to speak; the construction begins at the top and then adds lower and lower levels. Finally, by the semantical rules, the lowest level is anchored at the solid ground of the observable facts." Carnap, 65.

61. Oswald Veblen and John Wesley Young, *Projective Geometry* (Boston: Ginn and Company, 1910).

62. Owens, "Mathematics at War: Warren Weaver and the Applied Mathematics Panel 1942–1945."

63. Weaver began advocating for the development of applied mathematics as early as 1940. In a letter to George Birkhoff, Weaver stressed the importance of promoting applied mathematics, but was clear that Harvard was not the right place for this endeavor. MIT, he suggested, would be a more appropriate institution.

64. Warren Weaver to Marshall Stone, December 6, 1943, RAMP, box 15, folder "Correspondence 1944."

65. Marshall Stone to Warren Weaver, January 12, 1944, RAMP, box 15, folder "Correspondence 1944."

66. Stone to Weaver, January 12, 1944.

67. Stone to Weaver, January 12, 1944.

68. The literature on game theory is vast. See Paul Erickson, *The World the Game Theorists Made* (Chicago: University of Chicago Press, 2015); Robert J. Leonard, "Creating a Context for Game Theory," in *Toward a History of Game Theory*, ed. E. Roy Weintraub (Durham, NC: Duke University Press, 1992), 29–76; Robert Leonard, *Von Neumann, Morgenstern, and the Creation of Game Theory: From Chess to Social Science, 1900–1960* (New York: Cambridge University Press, 2010); Philip Mirowski, *Machine Dreams: Economics Becomes a Cyborg Science* (New York: Cambridge University Press, 2002).

69. For a comprehensive history of the phenomenon of game theory and its reach beyond the initial confines of economic theory, see Erickson, *The World the Game Theorists Made*.

70. Erickson et al., *How Reason Almost Lost*.

71. S. M. Amadae, *Rationalizing Capitalist Democracy: The Cold War Origins of Rational Choice Liberalism* (Chicago: University of Chicago Press, 2003), 6.

72. John von Neumann and Oskar Morgenstern, *Theory of Games and Economic Behavior*, commemorative ed. (Princeton, NJ: Princeton University Press, 2007 [1944]), 4.

73. Giambattista Formica, "Von Neumann's Methodology of Science: From Incompleteness Theorems to Later Foundational Reflections," *Perspectives on Science* 18, no. 4 (2010): 480–99.

74. von Neumann and Morgenstern, *Theory of Games*, 7.

75. von Neumann and Morgenstern, 74. Emphasis added.

76. von Neumann and Morgenstern, 28. Emphasis added.

77. Claude Elwood Shannon, "A Mathematical Theory of Communication," *Bell System Technical Journal* 27 (October 1948): 379–423, 623–56.

78. Ten years earlier, while still a student at MIT, Shannon published "A Symbolic Analysis of Relay and Switching Circuits," in which the influence of postulational thinking is more clearly evident. Claude E. Shannon, "A Symbolic Analysis of Relay and Switching Circuits," *Transactions of the American Institute of Electrical Engineers* 57, no. 12 (1938): 713–23.

79. Fritz Joachim Weyl, *Report on A Survey of Training and Research in Applied Mathematics in the United States* (Philadelphia: Society for Industrial and Applied Mathematics, 1956), 1.

80. Weyl, *Report on a Survey*, 11.

81. Weyl, 12.

82. Weyl, 10.

83. Weyl, 21.

84. Weyl, 26.

85. Weyl, 26.

86. Richard S. Burington, "On the Nature of Applied Mathematics," *American Mathematical Monthly* 56, no. 4 (1949): 222.

87. Burington, "Nature of Applied Mathematics," 222.

88. Burington even went on to assert that geometry existed simultaneously as a physical study and as pure mathematics.

89. Burington, "Nature of Applied Mathematics."

90. Griffith Evans, "Introductory Remarks on Applied Mathematics in the Traditional Departmental Structure," in *Proceedings of a Conference on Training in Applied Mathematics: Sponsored by the American Mathematical Society and by the National Research Council*, ed. Fritz Joachim Weyl (Washington, DC: National Academies Press, 1953), 12.

91. Evans, "Introductory Remarks," 12.

92. Weyl, *Proceedings of a Conference*, 23.

93. William Prager captured the sentiment perfectly when he remarked during his talk, "At a meeting such as this, one hears almost as many definitions of applied mathematics as there are speakers." William Prager, "The Graduate Division at Brown University," in Weyl, *Proceedings of a Conference*, 35.

94. John W. Tukey, "The Teaching of Concrete Mathematics," *American Mathematical Monthly* 65, no. 1 (January 1958): 1.

95. Tukey, "Teaching of Concrete Mathematics," 7.

96. Tukey, 7.

Chapter Three

1. Claude Lévi-Strauss, "The Mathematics of Man," *International Social Science Bulletin* 6, no. 4 (1954): 584.

2. Lévi-Strauss, "Mathematics of Man," 585.

3. David Aubin argued that Bourbaki served as a cultural connector for French structuralists. David Aubin, "The Withering Immortality of Nicolas Bourbaki: A Cultural Connector at the Confluence of Mathematics, Structuralism, and the Oulipo in France," *Science in Context* 10, no. 2 (1997): 297–342.

4. Lévi-Strauss, "Mathematics of Man," 586.

5. Lévi-Strauss, 586.

6. Kemeny explained that the classical definition of mathematics as the study of number and space did not sufficiently define modern mathematical research. Kemeny wrote that the appeal of mathematics to social scientists was that it offered a "convenient form in which to formulate scientific theories," and "force[d] that theoretician in various sciences to formulate his hypothesis in a precise and unambiguous form." Kemeny then proceeded to offer four concrete examples of how nonnumerical mathematics could be applied to social scientific research, in political science, communication networks, kinship theory, and decision theory. John G. Kemeny, "Mathematics without Numbers," *Daedalus* 88, no. 4 (1959): 577–91.

7. Theodore M. Porter, *Trust in Numbers: The Pursuit of Objectivity in Science and Public Life* (Princeton, NJ: Princeton University Press, 1996).

8. On postwar cybernetics, see Steve Joshua Heims, *Constructing a Social Science for Postwar America: The Cybernetics Group, 1946–1953* (Cambridge, MA: MIT Press, 1993); Ronald Kline, *The Cybernetics Moment: Or Why We Call Our Age the Information Age* (Baltimore: John Hopkins University Press, 2015); Orit Halpern, *Beautiful Data: A History of Vision and Reason since 1945* (Durham, NC: Duke University Press, 2015).

9. Dorothy Ross, "Changing Contours of the Social Science Disciplines," in *The Cambridge History of Science* ed. Theodore M. Porter and Dorothy Ross, vol. 7, *The Modern Social Sciences* (Cambridge: Cambridge University Press, 2003), 203–37; Theodore M. Porter and Dorothy Ross, ed., *The Cambridge History of Science*, vol. 7, *The Modern Social Sciences* (Cambridge University Press, 2003).

10. Joel Isaac, "Tangled Loops: Theory, History, and the Human Sciences in Modern America," *Modern Intellectual History* 6, no. 2 (2009): 409. Porter concurs with this view, noting that "the economic writings of physicists and engineers, at least up to the 1930s, suggests that the ambitions of scientists have been more closely allied with ideals of quantification and control than with abstract mathematical formulation. Measurement was not simply a link to theory, but a technology for managing events and an ethic that structured and gave meaning to scientific practice." Theodore M. Porter, *Trust in Numbers: The Pursuit of Objectivity in Science and Public Life* (Princeton, NJ: Princeton University Press, 1996), 72.

11. Isaac, "Tangled Loops," 410.

12. Clark Leonard Hull et al., *Mathematico-Deductive Theory of Rote Learning: A Study in Scientific Methodology* (New Haven, CT: Yale University Press, 1940).

13. Hull's work was not the first to extend axiomatic theory beyond mathematics and physics. In 1937, J. H. Woodger published *The Axiomatic Method in Biology*, in which he sought to place biological theory on logical grounds. Woodger, a British philosopher, was influenced by Russell and Whitehead's *Principia Mathematica*. Hull writes in the introduction to his book that discussions with Woodger convinced him of the importance of symbolic logic.

14. Hull et al., *Mathematico-Deductive Theory*, 10.

15. Gregory Bateson, "Reviewed Work: Mathematico-Deductive Theory of Rote Learning; A Study in Scientific Methodology," *American Anthropologist* New Series 43, no. 1 (1941): 116–18. Emphasis added.

16. The political exigencies of the Cold War made the promise of objective, value-free, and empirically verifiable theories increasingly appealing to funding agencies and scientists fearing ideological persecution in the age of McCarthyism. Mark Solovey, *Shaky Foundations: The Politics-Patronage-Social Science Nexus in Cold War America* (New Brunswick, NJ: Rutgers University Press, 2013).

17. A notable exception is Roy Weintraub's *How Economics Became a Mathematical Science* (Durham, NC: Duke University Press, 2002), which carefully traces the influence of Bourbaki axiomatics on economics.

18. Hunter Heyck, *Age of System: Understanding the Development of Modern Social Science* (Baltimore: Johns Hopkins University Press, 2015), 19.

19. Heyck, *Age of System*, 35.

20. Heyck also writes, "The adoption of a behavioral-functional approach was a critical step on the road to mathematization. Behavioral functional analysis did not lead inevitably to mathematical analysis (witness Talcott Parsons), but it did spur the development of mathematical social science in a couple of ways." Again, I do not mean to suggest that the behavioral-functional approach was simply the outcome of axiomatic thinking. Rather, I suspect that the two were mutually constitutive. As such, Parsons's work is instructive. While it was not mathematical, it subscribed to some of the main tenets of axiomatic analysis.

21. George Steinmetz, ed., *The Politics of Method in The Human Sciences: Positivism and Its Epistemological Others* (Durham, NC: Duke University Press, 2005).

22. Tony Lawson's contribution is the one exception, as Lawson emphasizes the influence of mathematical formalism, but it is limited to economics. George Steinmetz, *Politics of Method*; Tony Lawson, "Economic and Critical Realism: A Perspective on Modern Economics," in Steinmetz, *Politics of Method*, 366–92.

23. More than two hundred individuals applied for admission, out of which only fifty were accepted. A third of the participants had already obtained their PhD degree, and about half came from psychology.

24. Four years earlier, in 1949, another symposium on mathematical training for social scientists took place in Boulder, Colorado. This earlier symposium was organized as part of the summer meeting of the Econometric Society and was sponsored in collaboration with the Institute of Mathematical Statistics and the Mathematical Association of America. In the aftermath of the symposium, which was organized and chaired by Jacob Marschak, a committee was established in the hopes of fostering such endeavors (it included Madow, T. W. Anderson, Frank L. Griffin, Leonid Hurwicz, Marschak, and E. P. Northrop). In a report on the committee's recommendations in *Econometrica*, mathematician Frank Griffin singled out seven areas of research the committee identified as necessary (logarithms, graphs, interpolation, equations and forms of curves, probability, differential and integral calculus, and curve fitting). In addition, Griffin reported, the committee "urge[d] that special efforts should be made to acquaint the student with mathematical expression and reasoning as a mode of thought; and to cultivate a critical attitude in scrutinizing assumptions and the uses made of assumptions." The committee's report, thus, can be seen as an inflection point. On the one hand the mathematics deemed relevant did not yet reflect the theoretical reorientation in the field, but the emphasis on mathematical thinking, as opposed to technique, was already clearly present in it. The committee also distinguished between students in economics and in other social sciences. The former would require additional mathematical training, but not necessarily the latter. Yet, even for these other students, Griffin reported, some "elementary matrix theory" was recommended. Finally,

echoing Veblen almost perfectly, Griffin added, "I would, however, include a brief discussion of alternative postulational systems which give rise to rival systems of doctrine, with convenient illustrations in the non-Euclidean geometries. Such a discussion is an eye-opener for virtually all students and is a powerful antidote for dogmatic tendencies." "The Mathematical Training of Social Scientists, Report of the Boulder Symposium," *Econometrica* 18, no. 2 (1950): 193–205.

25. R. R. Bush et al., "Mathematics for Social Scientists," *American Mathematical Monthly* 61, no. 8 (1954): 554.

26. Bush et al., 558.

27. Bush et al., 558.

28. Bush et al., 558.

29. Bush et al., 555.

30. Bush et al., 561.

31. Robert R. Bush, Robert P. Abelson, and Ray Hyman, *Mathematics for Psychologists: Examples and Problems* (New York: Social Science Research Council, 1956).

32. While to the authors this is something to be admired about the work, it was one the main critiques of the New Math. Christopher J. Phillips, *The New Math: A Political History* (Chicago: University of Chicago Press, 2014). Richard Brandon Kershner and Lee Roy Wilcox, *The Anatomy of Mathematics* (New York: Ronald Press, 1950), v.

33. Kershner and Wilcox, *Anatomy of Mathematics*, 9.

34. Robert L. Davis, "Introduction to 'Decision Processes,'" in *Decision Processes*, ed. R. M. Thrall, C. H. Coombs, and R. L. Davis (New York: John Wiley & Sons, 1954), 3.

35. R. M. Thrall, C. H. Coombs, and R. L. Davis, eds. *Decision Processes* (New York: John Wiley & Sons, 1954), v.

36. Erickson explains, for example, that the Michigan seminar was considered a failure, as the participants at the conference did not necessarily cohere around a similar approach. Perhaps the only thing that united them was a shared orientation when it came to the applicability of mathematics. Paul Erickson, *The World the Game Theorists Made* (Chicago: University of Chicago Press, 2015).

37. Wilder, as discussed in chapter 1, was influential in orienting Steenrod to mathematics. At the University of Michigan, Wilder for many years taught a course on foundation that used Moore's theory. The first chapter of Wilder's book, which resulted from the course, is dedicated to a prolonged discussion of the axiomatic method. Raymond L. Wilder, *Introduction to the Foundations of Mathematics* (New York: John Wiley & Sons, 1952).

38. C. H. Coombs, Howard Raiffa, and R. M. Thrall, "Mathematical Models and Measurement Theory," in Thrall et al., *Decision Processes*, 26.

39. Howard Raiffa and Stephen E. Fienberg, "The Early Statistical Years: 1947–1967; a Conversation with Howard Raiffa," *Statistical Science* 23, no. 1 (2008): 136–49.

40. Angela M. O'Rand, "Mathematizing Social Science in the 1950s: The Early Development and Diffusion of Game Theory," in *Toward a History of Game Theory*, ed. E. Roy Weintraub (Durham, NC: Duke University Press, 1992), 189.

41. R. Duncan Luce and Howard Raiffa, *Games and Decisions: Introduction and Critical Survey* (New York: John Wiley & Sons, 1957), viii.

42. Luce and Raiffa, *Games and Decisions*, viii.

43. James S. Coleman, "Columbia in the 1950s," in *Authors of Their Own Lives: Intellectual Autobiographies by Twenty American Sociologists*, ed. Bennett M. Berger (Berkeley: University of California Press, 1990), 86.

44. Another publication by the Columbia Bureau of Applied Social Research in 1960 was entitled *Mathematical Thinking in the Measurement of Behavior*.

45. Paul Felix Lazarsfeld, ed., *Mathematical Thinking in the Social Sciences* (Glencoe, IL: Free Press, 1954), 3.

46. O'Rand, "Mathematizing Social Science."

47. Lazarsfeld, *Mathematical Thinking*, 5.

48. Kurt Lewin, *Field Theory in Social Science: Selected Theoretical Papers*, ed. Dorwin Cartwright (New York: Harper & Brothers, 1951), 3.

49. Theodora Dryer, "Designing Certainty: The Rise of Algorithmic Computing in an Age of Anxiety 1920–1970" (PhD diss., University of California San Diego, 2019).

50. Hermann Weyl, "A Half Century of Mathematics," *American Mathematical Monthly* 58, no. 8 (1951): 523.

51. Weyl, "Half Century of Mathematics," 425.

52. Weyl, "Half Century of Mathematics," 524.

53. This view was in no way limited to Weyl, but was taken as a truism among mathematicians. For example, in "The Architecture of Mathematics," an article attributed to Nicolas Bourbaki, the writer remarked, "It should be clear from what precedes that its [the axiomatic method] most striking feature is the effect of considerable economy of thought. The 'structures' are tools for the mathematicians; as soon as he has recognized among the elements, which he is studying, relations which satisfy the axioms of a known type, he has at his disposal immediately the entire arsenal of general theorems which belong to the structures of that type." Nicholas Bourbaki, "The Architecture of Mathematics," *American Mathematical Monthly* 57, no. 4 (1950): 221–32. Spellings of Bourbaki's first name varied between publications.

54. James S. Coleman, "An Expository Analysis of Some of Rashevsky's Social Behavior Models," in Lazarsfeld, *Mathematical Thinking*, 116.

55. Coleman, "Expository Analysis," 116.

56. Lazarsfeld, *Mathematical Thinking*, 8.

57. Frank Harary and R. Z. Norman, "Graph Theory as a Mathematical Model in Social Science," *Institute for Social Research*, 1953. Three years later, Harary would extend his use of directed graphs outside of psychology and sociology when

he offered a structural analysis of the situation in the Middle East. Frank Harary, "A Structural Analysis of the Situation in the Middle East in 1956," *Journal of Conflict Resolution* 5, no. 2 (1961): 167–78.

58. A decade later, Harary and Norman copublished, with psychologist Dorwin Cartwright, a book titled *Structural Models: An Introduction to the Theory of Directed Graphs*. The book, which was directed toward social scientists, opened with an axiomatic treatment of directed graphs. As the authors explained in the introduction, the presentation was used to demonstrate the logical basis of theory and the "association between graphs and relations." The presentation could also, the authors explained, "serve as an introduction for the student interested in the role of axiomatics in formal theories and in the general nature of mathematical models." Frank Harary, Robert Z. Norman, and Dorwin Cartwright, *Structural Models: An Introduction to the Theory of Directed Graphs* (New York: John Wiley & Sons, 1965), vi.

59. Weyl, "Half Century of Mathematics," 524.

60. Lazarsfeld, *Mathematical Thinking*, 5.

61. Hunter Crowther-Heyck, *Herbert A. Simon: The Bounds of Reason in Modern America* (Baltimore: Johns Hopkins University Press, 2005).

62. Herbert A. Simon, "Review: Theory of Games and Economic Behavior," *American Journal of Sociology* 50, no. 6 (May 1945): 559.

63. Simon, however, does insist in the preface that knowledge of calculus is important for social scientists. Herbert Simon, *Models of Man: Social and Rational* (New York: John Wiley and Sons, 1957), 89.

64. Simon, *Models of Man*, 99.

65. Simon, 114.

66. Simon, 114.

67. Simon, 114.

68. Simon, 114.

69. George A. Miller, "Applications of Mathematics in Social Psychological Research," *Items* 11, no. 4 (December 1957): 42.

70. "There is always a strong temptation to mistake the theory—especially a successful theory—for the actuality and to define the limits of a natural phenomenon in terms of the limits of our understanding of it." Miller, "Applications of Mathematics," 42.

71. Miller, "Applications of Mathematics," 44. Emphasis added.

72. Sonja M. Amadae, *Rationalizing Capitalist Democracy: The Cold War Origins of Rational Choice Liberalism* (Chicago: University of Chicago Press, 2003), 84–85.

73. Kenneth J. Arrow, *Social Choice and Individual Values* (New York: Wiley, 1951), 87.

74. Kenneth J. Arrow, "Mathematical Models in the Social Sciences," *General Systems: Yearbook of the Society for the Advancement of General Systems Theory* 1 (1956): 29. My emphasis.

75. Arrow, "Mathematical Models."

76. Anthony Downs, *An Economic Theory of Democracy* (New York: Harper & Row, 1957), 296–97.

77. Downs, *Economic Theory of Democracy*, 8.

78. Downs, 34.

79. James M. Buchanan and Gordon Tullock, *The Calculus of Consent: Logical Foundations of Constitutional Democracy* (Indianapolis: Liberty Fund, 1990), 17.

80. Amadae, *Rationalizing Capitalist Democracy*, chap. 3.

81. In a positive review of the book, Robert McGinnis noted, "Familiarity with (strict) monotone functions, and with sums of such functions, would suffice to convince the reader that the air of generality of this study is artificial." Remarking on the analysis in chapter 6 in the book, McGinnis writes that the authors introduce two functions, one that they call an "external cost function," and the other, "decision making costs." He then adds, "The net cost function is taken as the simple sum of these two functions and evidently is assumed to have a unique zero derivative point. All of this could be nonsense, either substantively or mathematically." Robert McGinnis, "Reviewed Work(s): The Calculus of Consent: Logical Foundations of Constitutional Democracy by James M. Buchanan and Gordon Tullock," *Annals of the American Academy of Political and Social Science* 346 (1963): 188–89.

82. Kenneth J. Arrow, "Tullock and an Existence Theorem," *Public Choice* 6 (1969): 105.

83. Buchanan and Tullock, *Calculus of Consent*, 296.

84. Buchanan and Tullock, 296.

85. Martin Diamond, "Reviewed Work(s): An Economic Theory of Democracy by Anthony Downs," *Journal of Political Economy* 67, no. 2 (1959): 208–11.

86. Hayward Rogers, "Some Methodological Difficulties in Anthony Downs's an Economic Theory of Democracy," *American Political Science Review* 53, no. 2 (1959): 484.

87. Rogers, "Some Methodological Difficulties," 485.

88. Anthony Downs, "Dr. Rogers's Methodological Difficulties: A Reply to His Critical Note," *American Political Science Review* 53, no. 4 (1959): 1095.

89. Buchanan and Tullock, *Calculus of Consent*, 295.

90. Joel Isaac, "Tool Shock: Technique and Epistemology in the Postwar Social Sciences," *History of Political Economy* 42 (2010): 135.

91. Davis, "Introduction to 'Decision Processes,'" 4.

92. Davis, "Introduction to 'Decision Processes,'" 5.

93. Luce and Raiffa, *Games and Decisions*, 36. My emphasis.

94. Luce and Raiffa, *Games and Decisions*, 37.

95. Downs, *Economic Theory*, 14.

96. Here, Downs footnotes Milton Friedman, "The Methodology of Positive Economics." Milton Friedman, "The Methodology of Positive Economics," in *Essays in Positive Economics* (Chicago: University of Chicago Press, 1953), 3–46.

97. Downs explains that this proposition is based on the hypothesis that political parties in a democracy plan their policies so as to maximize votes. Downs, *Economic Theory*, 297.

98. Buchanan and Tullock, *Calculus of Consent*, 17.

99. Emily Hauptman has argued that the meaning of positivism for Riker is defined through its main opposition. Thus, in the 1950s and 1960s it was defined against normative theory, and in the 1970s it was defined against empirical inductive theory. Emily Hauptmann, "Political Science/Political Theory Defining 'Theory' in Postwar Political Science," in *The Politics of Method in the Human Sciences: Positivism and Its Epistemological Others*, ed. George Steinmetz (Durham, NC: Duke University Press, 2005), 207–32.

100. Riker and Ordeshook, *Positive Political Theory*, xi.

101. Anatol Rapoport, "Various Meanings of 'Theory,'" *American Political Science Review* 52, no. 4 (1958): 979.

102. Rapoport, 980.

103. Anatol Rapoport, "Uses and Limitations of Mathematical Models in Social Science," in *Symposium on Sociological Theory*, ed. Llewellyn Gross (New York: Harper & Row, 1959), 371. The symposium included several talks on axiomatics. Gross, *Symposium on Sociological Theory*.

104. As Debora Hammond explains, Rapoport had a different definition of normative than was common at the time. For example, he writes that Galileo's theory was normative because it outlined how bodies "ought to fall under idealized conditions." Debora Hammond, *The Science of Synthesis: Exploring the Social Implications of General Systems Theory* (Boulder: University Press of Colorado, 2003).

105. Rapoport, "Various Meanings of 'Theory,'" 987.

106. Rapoport, 987.

107. Rapoport, 983.

108. James S. Coleman, *Introduction to Mathematical Sociology* (New York: Free Press of Glencoe, 1964), 34.

109. Coleman, *Introduction to Mathematical Sociology*, 37.

110. Coleman, 50.

111. Buckminster R. Fuller, "Conceptuality and Fundamental Structures," in *Structure in Art and in Science*, ed. György Kepes (New York: G. Braziller, 1965), 68.

112. Theodora Vardouli, "Graphing Theory: New Mathematics, Design, and the Participatory Turn" (PhD diss., Massachusetts Institute of Technology, 2017), 18.

113. Vardouli, "Graphing Theory."

114. I expand on this idea in Alma Steingart, "The Axiomatic Aesthetic," in *Computer Architectures: Constructing the Common Ground*, ed. Theodora Vardouli and Olga Touloumi (London: Routledge, 2019).

115. Yona Friedman, *Toward a Scientific Architecture*, trans. Cynthia Lang (Cambridge, MA: MIT Press, 1975), 20–25.

116. Friedman, *Toward a Scientific Architecture*, 25–26.

117. Friedman, *Toward a Scientific Architecture*, 25–26.

Chapter Four

1. The interaction was reported in the *New Yorker*'s About Town section. "Mathematicians," *The New Yorker*, November 16, 1963.

2. Both artists and critics have questioned the accuracy of using "Abstract Expressionism" to denote a varied collection of artists including Pollock, Rothko, Newman, Motherwell, and others. For historians of art, if something united these various artists, it was their shared desire to "translate private feelings and emotions directly onto the material field of the canvas—without the mediation of figurative content." Hal Foster et al., *Art Since 1900: Modernism Antimodernism Postmodernism* (London: Thames & Hudson, 2016), 404.

3. Dorothy Seiberling, "Varied Art of Four Pioneers: Analogies with Nature Help Explain Abstract-Expressionist Work," *Life*, November 16, 1959.

4. When in 1951 E. R. Lorch reviewed Halmos's *Introduction to Hilbert Space and the Theory of Spectral Multiplicity*, he noted, "In one hundred and nine well-packed pages one finds an exposition which is always free, proofs which are sophisticated, and a choice of subject matter which is certainly timely. Some of the vineyard workers will say that P. R. Halmos has become addicted to the delights of writing expository tracts. Judging from recent results one can only wish him continued indulgence in this attractive vice." E. R. Lorch, "Review: Introduction to *Hilbert Space and the Theory of Spectral Multiplicity*," *Bulletin of the American Mathematical Society* (May 1952): 412.

5. "The methodologist (pure mathematician) feels the need to justify pure mathematics exactly as little as the musician feels the need to justify music. Do practical men, the men who meet payrolls, demand only practical music—soothing jazz to make an assembly line worker turn nuts quicker, or stirring marches to make a solider kill with more enthusiasm? No, surely none of us believes in that kind of justification." Paul R. Halmos, "Mathematics as a Creative Art," *American Scientist* 56, no. 4 (1968): 387.

6. Halmos, "Creative Art," 388.

7. Halmos, "Creative Art," 388.

8. Halmos, "Creative Art," 389.

9. Halmos, "Creative Art," 389.

10. Max Kozloff, "American Painting during the Cold War," *Artforum* 11 (May 1973): 43–54; Eva Cockcroft, "Abstract Expressionism, Weapon of the Cold War," *Artforum* 12 (June 1974): 39–41; Michael Kimmelman, "Revisiting the Revisionists: The Modern, Its Critics, and the Cold War," in *The Museum of Modern Art at Mid-Century: At Home and Abroad, Studies in Modern Art* 4 (New York: MoMA,

1994): 38–55; Robert Burstow, "The Limits of Modernist Art as a 'Weapon of the Cold War': Reassessing the Unknown Patron of the Monument to the Unknown Political Prisoner," *Oxford Art Journal* 20, no. 1 (1997): 68–80; Irving Sandler, "Abstract Expressionism and the Cold War," in *Abstract Expressionism and the American Experience: A Reevaluation* (New York: Hard Press Editions, 2009), 173–96.

11. Serge Guilbaut, *How New York Stole the Idea of Modern Art* (Chicago: University of Chicago Press, 1985). It is worth noting that mathematicians similarly pointed out that the apolitical nature of mathematical work could help foster international exchange. Michael J. Barany, "Fellow Travelers and Traveling Fellows: The Intercontinental Shaping of Modern Mathematics in Mid-Twentieth Century Latin America," *Historical Studies in the Natural Sciences* 46, no. 5 (2016): 669–709; Brit Shields, "Mathematics, Peace, and the Cold War: Scientific Diplomacy and Richard Courant's Scientific Identity," *Historical Studies in the Natural Sciences* 46, no. 5 (2016): 556–91.

12. T. J. Clark has argued that, even if accurate, such a historical account of Abstract Expressionism does not answer the questions "To what extent was the meeting of class and art practice in the later 1940s more than just contingent? To what extent does Abstract Expressionism really belong, at the deepest level—the level of language, of procedure, or presuppositions about world making—to the bourgeoisie who paid for it and took it on their travels?" In other words, the emphasis on the Cold War might explain why the work was co-opted, but not the motivations and subjective experiences of the artists themselves. T. J. Clark, "In Defense of Abstract Expressionism," *October* 69 (1994): 26.

13. George L. K. Morris et al., "What Abstract Art Means to Me," *Bulletin of the Museum of Modern Art* 18, no. 3 (Spring 1951): 12.

14. According to Leah Dickerman, one of the main characteristics of the early-twentieth-century emergence of abstraction in art was its associated written material. "Abstract pictures rarely, if ever, exist in isolation. The makers of early abstract pictures and their allies did not let them stand alone, but—perhaps in compensation for shared anxieties about how its meaning might be established—sent them out into the world accompanied by a torrent of words: titles, manifestos, statements of principles, performative declamations, discursive catalogues, explanatory lectures, and critical writing by allies." Yve-Alain Bois et al., "Abstraction, 1910–1925: Eight Statements," *October* 143 (Winter 2013): 4–5. By midcentury, artists still felt the need to explain the "meaning" of their abstraction. In 1951, the Museum of Modern Art, for example, presented a symposium on abstract art and invited artists to offer their views on the topic "What abstract art means to me." The artists included George L. K. Morris, Willem de Kooning, Alexander Calder, Fritz Glarner, Robert Motherwell, and Stuart Davis. This is not to say that these artists agreed on a coherent meaning for abstraction in art, but rather that the question was alive and well. Morris et al., "What Abstract Art Means."

15. Historian of art Yve-Alain Bois identified four main strategies or models that abstract artists adopted to "justify (for themselves and for their audience) what they were doing." As Bois noted, abstract artists' freedom from representation brought on a new demand for explanations of their work. The first model Bois identified was the Romantic ideal, which conceived of artists as operating in complete freedom. The second model, which he termed "iconological," asserted that abstract art depicted abstract concepts, such as light, color, and so on. The third model, "compositional," approached painting as a manifestation of "an extremely complex system of thought," as exemplified by the work of Piet Mondrian. The last model was "non-composition," which stood in direct opposition to the previous three models and consisted of a "programmatic insistence on the non-agency of the artist." The last model, which Bois went on to subdivide into six subcategories, was by far the most common in twentieth-century abstract art. Bois et al., "Abstraction, 1910–1925."

16. The relation between art and mathematics goes as far back as ancient Greece, and is most famously associated with the Pythagorean tradition that linked some consonant quality of musical intervals with whole numbers. Indeed, music in particular has had a long-standing bond with mathematics (as its inclusion among the quadrivium subjects clearly indicates), and over the centuries numerous mathematicians have studied how mathematical ideas, from logarithms to combinatorics, can inform musical theory. Starting in the Renaissance, mathematics and the visual arts also became wedded to one another through the growth of linear perspective. In literature as well, mathematical ideas have been present for centuries. In other words, there was nothing new per se in claiming a relation between mathematics and the arts in the postwar period. Rather, what distinguished the postwar moment was the *type* of relation between the two that mathematicians wished to highlight. The literature on the relation between art and mathematics is too vast. For an overall analysis of this history starting in prehistory and continuing to the present, see Lynn Gamwell, *Mathematics and Art: A Cultural History* (Princeton, NJ: Princeton University Press, 2015). On the relation between mathematics and music, see Gerard Assayag, Hans George Feichtinger, and Jose Francisco Rodrigues, ed., *Mathematics and Music: A Diderot Mathematical Forum* (Berlin: Springer-Verlag, 2002); John Fauvel, Raymond Flood, and Robin Wilson, ed., *Music and Mathematics: From Pythagoras to Fractals* (New York: Oxford University Press, 2003); Penelope Gouk, *Music, Science, and Natural Magic in Seventeenth-Century England* (New Haven, CT: Yale University Press, 1999); Peter Pesic, "Hearing the Irrational: Music and the Development of the Modern Concept of Number," *Isis* 101, no. 3 (2010): 501–30. On the relation between mathematics and the visual arts, see Martin Kemp, *The Science of Art: Optical Themes in Western Art from Brunelleschi to Seurat* (New Haven, CT: Yale University Press, 1990); Judith Veronica Field, *The Invention of Infinity: Mathematics and Art in the Renaissance* (New York: Oxford University Press, 1997). Finally, on the relation between

mathematics and literature, see Robert Tubbs, Alice Jenkins, and Nina Engelhardt, ed., *The Palgrave Handbook of Literature and Mathematics* (Cham, Switzerland: Palgrave Macmillan, 2021).

17. On the relation between artistic modernism and mathematical modernism, see Linda Dalrymple Henderson, *The Fourth Dimension and Non-Euclidean Geometry in Modern Art* (Cambridge, MA: MIT Press, 2013); Andrea K. Henderson, *Algebraic Art: Mathematical Formalism and Victorian Culture* (Oxford: Oxford University Press, 2018). Nina Engelhardt, *Modernism, Fiction and Mathematics* (Edinburgh: Edinburgh University Press, 2018).

18. More recently, some historians of mathematics have argued that mathematics should be understood as a narrative practice. Amir R. Alexander, "Tragic Mathematics: Romantic Narratives and the Refounding of Mathematics in the Early Nineteenth Century," *Isis* 97, no. 4 (December 2006): 714–26; Apostolos K. Doxiadis and Barry Mazur, *Circles Disturbed: The Interplay of Mathematics and Narrative* (Princeton, NJ: Princeton University Press, 2012).

19. Marston Morse, "Mathematics and the Maximum Scientific Effort in Total War," *Scientific Monthly* 56, no. 1 (1943): 55.

20. Morse, "Maximum Scientific Effort," 55.

21. As noted in chapter 5, Morse's worries did not materialize. In the academic environment, it was pure rather than applied mathematics that represented the lion's share of growth in the field.

22. Griffith Evans to Marshall Stone, September 24, 1941, MMP, box 6, folder "Evans."

23. Richard Courant to Ronald Kline, November 21, 1944, AMSR, box 29, folder 39.

24. Courant to Kline, November 21, 1944. By the 1950s, Courant's worry had changed. In the war's aftermath, Courant believed that it was applied and *not* pure mathematics that had not developed sufficiently. He attributed this fact to a long-held belief among mathematicians that pure mathematics was superior to applied mathematics.

25. Morse, "Maximum Scientific Effort."

26. Morse, 55.

27. Matthew L. Jones, *The Good Life in the Scientific Revolution: Descartes, Pascal, Leibniz, and the Cultivation of Virtue* (Chicago: University of Chicago Press, 2008).

28. Joan L. Richards, "'This Compendious Language': Mathematics in the World of Augustus De Morgan," *Isis* 102, no. 3 (2011): 510.

29. Lewis Pyenson, *Neohumanism and the Persistence of Pure Mathematics in Wilhelmian Germany* (Philadelphia: American Philosophical Society, 1983).

30. Christopher Phillips, *The New Math: A Political History* (Chicago: University of Chicago Press, 2015).

31. This sentiment was prevalent. In February 1945, a group of leading mathematicians met at Swarthmore to discuss the future of mathematical education,

which they believed had great shortcomings. In the short report they authored, they explained that "like language and music it [mathematics] is one of the primary manifestations of the free creative power of the human mind, and it is the universal organ for world-understanding through theoretical construction. Mathematics must therefore remain an essential element of the knowledge and abilities which we have to teach, of the culture we have to transmit to the next generation." The authors included Richard Courant, Arnold Dresden, J. R. Kline, Hans Rademacher, and Hermann Weyl. Undated, AMSR, box 30, folder 47.

32. Marston Morse, "Science and the Library," *College and Research Libraries* 10, no. 2 (April 1949): 152.

33. In other words, Morse was after "pure," not "basic," science. Rebecca S. Lowen, "The More Things Change . . . : Money, Power and the Professoriate," *History of Education Quarterly* 45, no. 3 (2005): 438–45.

34. Jessica Wang, "Physics, Emotion, and the Scientific Self: Merle Tuve's Cold War," *Historical Studies in the Natural Sciences* 42, no. 5 (November 2012): 341–88.

35. Wang, "Physics, Emotion," 372.

36. Here the growth of computer science is the obvious exception, as it did rely on substantial support.

37. I do not mean to suggest that the image of pure science Morse describes ever existed. Yet this nostalgic image of science served as motivation for Morse.

38. Marshall H. Stone, "Mathematics and the Future of Science," *Bulletin of the American Mathematical Society* 63, no. 2 (1957): 66.

39. Stone, "Mathematics and the Future," 66.

40. Alan T. Waterman, "The National Science Foundation Program in Mathematics," *Bulletin of the American Mathematical Society* 60, no. 3 (1954): 207–14.

41. Carl B. Boyer, "Mathematical Inutility and the Advance of Science," *Science* 130, no. 3366 (July 3, 1959): 22–25.

42. Jessica Wang, *American Science in an Age of Anxiety: Scientists, Anticommunism, and the Cold War* (Chapel Hill: University of North Carolina Press, 2000), 219–52.

43. Stone was not the only mathematician who wrote to Bronk in protest. Oswald Veblen wrote that he "was terribly disappointed at the responses made by Richards and [Bronk] on the subject of requiring oaths and affidavits." Veblen insisted that the intrusion of any requirement besides scientific competence on fellowship candidates would be a hindrance to science. But his objection was not only made on scientific grounds. Veblen believed that there was such a lack of oversight that an organization deemed fine today could turn dangerous tomorrow. Bronk replied by trying to reassure Veblen that he was committed to the freedom of scientists, insisting that he had to do whatever was possible to ensure the continuation of the fellowship program. Veblen must not have been satisfied. Five days after receiving Bronk's reply, he directed his complaint to none other than President Truman, protesting "that there [was] no judicial process by which an organization

[was] placed upon or removed from [the] list" of subversive organizations issued by the attorney general. Oswald Veblen to Detlev Bronk, June 15, 1949; Detlev Bronk to Oswald Veblen, June 25, 1949; Oswald Veblen to President of the United States, June 30, 1949, OVP, box 25, folder "Loyalty Problem and Academic Freedom, 1949."

44. Marshall H. Stone, "The AEC Loyalty Oath," *Science* 110, no. 2851 (August 19, 1949): 191–92.

45. Minersville School District v. Gobitis, 310 US 586 (1940).

46. Stone, "AEC Loyalty Oath," 191.

47. Irving Segal to Ronald Kline, August 9, 1949, AMSR, box 35, folder 133.

48. The history of the case is given in "Majority: Report of a Committee to Advise the Council in Relation to the Situation at the University of California," AMSR, box 36, folder 12.

49. After the instantiations of the loyalty oath, the university dismissed approximately thirty professors who refused to sign the oath. Ellen W. Schrecker, *No Ivory Tower: McCarthyism and the Universities* (Oxford: Oxford University Press, 1986), 117.

50. "Majority: Report of a Committee to Advise the Council in Relation to the Situation at the University of California," AMSR, box 36, folder 12.

51. In 1948 the society also weighed in on the case of Edward Condon, who was targeted by the House Un-American Activities Committee and accused of being a Communist sympathizer. The society condemned the actions of the committee: "These procedures will, if continued, enable congressional committees to usurp and abuse functions properly reserved under our Constitution to the executive and judicial branches of our government." On the Condon case, see Wang, *American Science*.

52. "Majority: Report of a Committee."

53. "Majority: Report of a Committee."

54. David W. Carroll, "The Regents Versus the Professors: Edward Tolman's Role in the California Loyalty Oath Controversy," *Journal of the History of the Behavioral Sciences* 48, no. 3 (Summer 2012): 218–35.

55. Adrian Albert, "Mathematics as a Profession," Youth Conference on the Atom, October 20, 1960, AAAP, box 2.

56. This information is taken from a letter Morse wrote in 1953 to Donald Adams. Marston Morse to Donald J. Adams, May 7, 1953, MMP, box 1, folder A. In a letter to Arnaud Denjoy in 1970, Morse wrote, "the poet, Robert Frost, was a friend of mine." It is not clear to me whether they were on friendly terms before the conference or only afterward. Marston Morse to Arnaud Denjoy, July 2, 1970, MMP, box 4, folder "Denjoy."

57. Hermann Weyl, *Symmetry* (Princeton, NJ: Princeton University Press, 1952).

58. Donald J. Adams, "Speaking of Books," *New York Times Book Review*, November 12, 1950.

59. Raoul Bott, "Marston Morse and His Mathematical Works," *Bulletin of the American Mathematical Society* 3, no. 3 (1980): 907–50.

60. Morse was not the first to make this point. See Jacques Hadamard, *The Psychology of Invention in the Mathematical Field* (New York: Dover Publications, 1954).

61. Robert Frost, "The Figure a Poem Makes," in *Collected Poems of Robert Frost* (New York: Holt, Rinehart, and Winston, 1939).

62. Marston Morse, "Mathematics and the Arts," *Yale Review* 40 (1951): 604–12.

63. Here I do not mean creators in the philosophical sense of whether mathematical concepts are created or discovered.

64. In 1965, mathematician Salomon Bochner read the entire history of mathematics through the lens of creation. "Strictly speaking, however, mathematics deals only with objects of its own imagery, which are internally conceived, internally created, and inwardly structured." Salomon Bochner, "Why Mathematics Grows," *Journal of the History of Ideas* 26, no. 1 (1965): 17–18.

65. Martin H. Krieger, trans. "A 1940 Letter of André Weil on Analogy in Mathematics," *Notices of the AMS* 52, no. 3 (2005): 341.

66. Norbert Wiener, *Ex-Prodigy: My Childhood and Youth* (Cambridge, MA: MIT Press, 1953), 212.

67. Herschel B. Chipp, ed., *Theories of Modern Art: A Source Book by Artists and Critics* (Berkeley: University of California Press, 1984), 536.

68. Chipp, *Theories of Modern Art*, 536.

69. Linda Dalrymple Henderson, "Editor's Introduction: I. Writing Modern Art and Science—An Overview; II. Cubism, Futurism, and Ether Physics in the Early Twentieth Century," *Science in Context* 17, no. 4 (2004): 423–66.

70. The other common denominator which scientific humanists emphasized was structure. Elsewhere, I have argued that while the concept of structure that artists employed was undoubtedly indebted to mathematics, it is somewhat ironic that artists turned to mathematical structure, as it was this emphasis on structuralism that turned mathematicians' eyes away from the world. Alma Steingart, "The Axiomatic Aesthetic," in *Computer Architectures: Constructing the Common Ground*, ed. Theodora Vardouli and Olga Touloumi (London: Routledge, 2019).

71. Four years before Morse gave his speech, another American mathematician, Archibald Henderson, made a similar claim in "Science and Art: An Approach to a New Synthesis." Henderson, like Morse, focused on the role of intuition and imagination. Interestingly, in arguing for a bond between art and science, Henderson offered Marston Morse as an example of "representatives of the unexpected, apparently inexplicable, composite of artist and scientist." The other "representatives" included Johann Wolfgang von Goethe, Oliver Wendell Holmes, W. B. Smith, and Albert Einstein. Archibald Henderson, "Science and Art: An Approach to a New Synthesis," *American Scientist* 34, No. 3 (1946): 460.

72. Morse, "Mathematics and the Arts," 612.

73. Morse, "Mathematics and the Arts," 612.

74. Marston Morse, "Mathematics and the Arts," *Bulletin of the Atomic Scientists* 9 (1959): 59.

75. Morse, "Mathematics and the Arts," 59.

76. National Research Council, *The Mathematical Sciences: A Report* (Washington, DC: National Academies Press, 1968), 220.

77. National Research Council, *Mathematical Sciences*, 220.

78. Jamie Cohen-Cole, *The Open Mind: Cold War Politics and the Sciences of Human Nature* (Chicago: University of Chicago Press, 2014).

79. Peter D. Lax, "Mathematics and Its Applications," *Mathematical Intelligencer* 8, no. 4 (1986): 14–17.

80. Lax, "Mathematics and Its Applications," 15.

81. Lax, like Halmos, suggested that "as art mathematics resemble[d] most closely painting." Both activities, he argued, were defined by a tension between relating to the world and searching for beauty. And in *both* activities, the best work was the one in which the tension between the two tendencies was the greatest. "The least satisfactory," he added, "are those works where one aspect predominated, as in genre painting or pure abstraction." Here "pure abstraction" refers to art, but could just as easily be applied to mathematics. Lax, "Mathematics and Its Applications," 15.

82. Von Neumann did not support this tendency. He believed mathematicians must continuously search for inspiration in the world around them. John von Neumann, "The Mathematician," in *The Works of the Mind*, ed. Robert B. Heywood (Chicago: University of Chicago Press, 1947), 180–96.

83. The value of the axiomatic study of geometry was "mainly epistemological." Leo Corry, *David Hilbert and the Axiomatization of Physics (1898–1918): From Grundlagen der Geometrie to Grundlagen der Physik* (New York: Springer Science+Business Media, 2004), 83.

84. David Hilbert and S. Cohn-Vossen, *Geometry and the Imagination* (New York: Chelsea Publishing Company, 1952), iii–iv.

85. Gamwell, *Mathematics and Art*.

86. Henri Poincaré, *Science and Method*, trans. Francis Maitland (New York: Thomas Nelson, 1914), 22.

87. Von Neumann, "The Mathematician."

88. Von Neumann, "The Mathematician," 195.

89. John Tukey to Lipman Bers, March 23, 1966, JWTP, series I, box 23, folder "National Research Council: Committee on Support of Research in the Mathematical Sciences."

90. Clement Greenberg, "Towards a Newer Laocoön," in *The Collected Essays and Criticism*, vol. 1, *Perceptions and Judgements 1939–1944* (Chicago: University of Chicago Press, 1986), 28.

91. Greenberg, "Towards a Newer Laocoön," 28.

92. Clement Greenberg, "The New Sculpture," in *Art and Culture: Critical Essays* (Boston: Beacon Press, 1989), 139.

93. Gamwell, *Mathematics and Art*, 153–54.

94. One can try to draw an even more direct line. According to Victor Erlich, the Russian Formalists were strongly influenced by the writing of Edmund Husserl, who, of course, began his philosophical investigation into the foundation of arithmetic and was also a close colleague of Hilbert. Victor Erlich, trans., *Russian Formalism: History–Doctrine* (New York: Mouton Publishers, 1955), 61–62.

95. On the relation between logic and literature, see Jeffrey Blevins and Daniel Williams, "Introduction: Logic and Literary Form," *Poetics Today* 41, no. 1 (2020): 1–36.

96. Gamwell, *Mathematics and Art*, 215.

97. Jameson, *A Singular Modernity*, 165.

98. Jameson, *A Singular Modernity*, 165.

99. Jameson, *A Singular Modernity*, 172.

100. Clement Greenberg, "The Case for Abstract Art," in *The Collected Essays and Criticism*, vol. 4, *Modernism with a Vengeance, 1957–1969*, ed. John O'Brian (Chicago: The University of Chicago Press, 1993), 75–84.

101. Greenberg, "Case for Abstract Art," 80.

102. Lax, "Mathematics and Its Applications," 15.

103. Morse, "Maximum Scientific Effort," 55.

104. Fritz Joachim Weyl, *Report on a Survey of Training and Research in Applied Mathematics* (Philadelphia: Society for Industrial and Applied Mathematics, 1956), 11–12.

105. Johannes Weissinger, "The Characteristic Features of Mathematical Thought," in *The Spirit and the Uses of the Mathematical Sciences*, ed. Thomas Lorie Saaty and Fritz Joachim Weyl (New York: McGraw-Hill, 1969), 18.

106. Theodore M. Porter, "Foreword: Positioning Social Science in Cold War America," in *Cold War Social Science: Knowledge Production, Liberal Democracy, and Human Nature*, ed. Mark Solovey and Hamilton Cravens (New York: Palgrave Macmillan, 2012), x.

107. Porter, "Foreword," ix.

108. Allen Tate, "Miss Emily and the Bibliographer," *American Scholar* 9, no. 4 (1940): 456.

109. Gerald Graff, *Literature against Itself: Literary Ideas in Modern Society* (Chicago: Ivan R. Dee, 1995), 129.

110. Caroline A. Jones, *Eyesight Alone: Clement Greenberg's Modernism and the Bureaucratization of the Senses* (Chicago: University of Chicago Press, 2005), 104.

111. Jones, *Eyesight Alone*, 60.

112. Foster et al., *Art Since 1900*, 380.

113. Greenberg's formalism, according to Caroline Jones, was influenced by logical positivists, who were in turn influenced by turn-of-the-century foundational concerns in mathematics.

114. Porter, "Foreword," ix.

115. Jones, *Eyesight Alone*, 64.

Chapter Five

1. Daniel J. Kevles, *The Physicists: The History of a Scientific Community in Modern America* (Cambridge, MA: Harvard University Press, 1995), 421.

2. Roger L. Geiger, *Research and Relevant Knowledge: American Research Universities since World War II* (New York: Oxford University Press, 1993); David Kaiser, "Cold War Requisitions, Scientific Manpower, and the Production of American Physicists after World War II," *Historical Studies in the Physical and Biological Sciences* 33, no. 1 (2002): 131–59; Kevles, *Physicists*.

3. R. C. Buck, "Relations with the Government," December 5, 1971, SMP, box 6.4/AAM 86–10/4, folder "AMS: Government Relations, Committee On—Part Two."

4. Buck, "Relations with the Government."

5. Alma Steingart, "Conditional Inequalities: American Pure and Applied Mathematics, 1940–1975" (PhD diss., Massachusetts Institute of Technology, 2013), chap. 5.

6. See, for example: Paul Forman, "Behind Quantum Electronics: National Security as Basis for Physical Research in the US, 1940–1960," *Historical Studies in the Physical Sciences* 18, no. 1 (1987); Daniel J. Kevles, "Cold War and Hot Physics: Science, Security and the American State, 1945–56," *Historical Studies in the Physical and Biological Sciences* 20, no. 2 (1990): 239–64; Chandra Mukerji, *A Fragile Power*; David H. DeVorkin, *Science with a Vengeance: How the Military Created the US Space Sciences after World War II* (New York: Springer-Verlag, 1992); Finn Aaserud, "Sputnik and the 'Princeton Three:' The National Security Laboratory That Was Not to Be," *Historical Studies in the Physical and Biological Sciences* 25, no. 2 (1995): 185–239; Mark Solovey, "Introduction: Science and the State during the Cold War: Blurred Boundaries and a Contested Legacy," *Social Studies of Science* 31, no. 2 (2001): 165–70; Jacob Darwin Hamblin, *Oceanographers and the Cold War: Disciples of Marine Science* (Seattle: University of Washington Press, 2005); Hunter Heyck and David Kaiser, "Introduction: New Perspectives on Science and the Cold War," *Isis* 101, no. 2 (2010): 362–66; Joel Isaac, "Introduction: The Human Sciences and Cold War America," *Journal of the History of the Behavioral Sciences* 47, no. 3 (2011): 225–31.

7. A related, yet distinct, concern has been the impact of McCarthyism on the nature of intellectual inquiry. A turn toward scientism and propensity for depoliticized research became a hallmark of the postwar social sciences, philosophy, and humanities. Stuart W. Leslie, *The Cold War and American Science: The Military-Industrial-Academic Complex at MIT and Stanford* (New York: Columbia University Press, 1993); Ellen W. Schrecker, *No Ivory Tower: McCarthyism and the Universities* (Oxford: Oxford University Press, 1986); Jessica Wang, *American Science in an Age of*

Anxiety: Scientists, Anticommunism, and the Cold War (Chapel Hill: University of North Carolina Press, 2000). The rise of area studies and analytic philosophy have been singled out, as direct outcomes of the Cold War ideological battles. John McCumber, *Time in the Ditch: American Philosophy and the McCarthy Era* (Evanston, IL: Northwestern University Press, 2001); George A. Reisch, *How the Cold War Transformed Philosophy of Science: To the Icy Slopes of Logic* (Cambridge: Cambridge University Press, 2005); David C. Engerman, *Know Your Enemy: The Rise and Fall of America's Soviet Experts* (New York: Oxford University Press, 2009).

8. Naomi Oreskes, *Science on a Mission: How Military Funding Shaped What We Do and Don't Know about the Ocean* (Chicago: University of Chicago Press, 2020).

9. Amy Dahan-Dalmédico, "An Image Conflict in Mathematics after 1945," in *Changing Images in Mathematics: From the French Revolution to the New Millennium*, ed. Umberto Bottazzini, and Amy Dahan-Dalmédico (New York: Routledge, 2001): 223–53.

10. In an interview in 1981, Paul Halmos lashed out against the postwar support structure: "I think we have been given too much money. I don't think mathematics needs to be supported . . . Mathematics gets along fine, thank you, without money." Halmos argued that mathematicians would continue to produce mathematical research without any external support. Moreover, he believed that federal support had a corrupting influence on the field:

> In the fifties and sixties, a lot of people went into mathematics for the wrong reasons, namely that it was glamorous, socially respected, and well-paying. The Russians fired Sputnik, the country became hysterical, and the NSF came along with professional, national policies. Anything and everything was tried; nothing was too much. We had to bribe people to come to mathematics classes to make it appear respectable, glamorous, and well-paying. So we did. One way we did it, for instance, was to use a completely dishonest pretense—the mission attitude towards mathematics. The way it worked was that I would propose a certain piece of research to do, I would get some money. That's so dishonest it sickens me. None of it was true! We got paid for doing research because the country wanted to spend money training mathematicians to help fight the Russians. (Donald J. Albers and Gerald L. Alexanderson, ed., *Mathematical People: Profiles and Interviews* [Boston: Birkhäuser, 2008], 127–28; see Chandra Mukerji, *A Fragile Power: Scientists and the State* [Princeton, NJ: Princeton University Press, 1989])

11. During a conference in 1953 on training in applied mathematics, representatives from various industries were invited to discuss potential employment for mathematicians in industry. However, all speakers spent more time discussing the

incompatibility of mathematical training for industrial work than describing its usefulness. E. C. Nelson from Hughes Aircraft Company, for example, noted that the abstract training of pure mathematicians might "explain the frequent occurrence of situations in which mathematician members of research teams become preoccupied with problems that the scientist members regard as unimportant." E. C. Nelson, "The Role of the Mathematician in Industrial Laboratories," in *Proceedings of a Conference on Training in Applied Mathematics: Sponsored by the American Mathematical Society and by the National Research Council*, ed. Fritz Joachim Weyl (Washington, DC: National Academies Press, 1953), 95.

12. Fritz Joachim Weyl, *Report on A Survey of Training and Research in Applied Mathematics in the United States* (Philadelphia: Society for Industrial and Applied Mathematics, 1956), 3.

13. As I have shown elsewhere, the growth of applied mathematics also points to the limited influence of external funding alone on the development of new scientific knowledge. It is precisely the close association of the field with the military that accounted for its failure to thrive in American universities. Steingart, "Conditional Inequalities," chap. 2.

14. Steingart, "Conditional Inequalities," chap. 4.

15. National Research Council, *The Mathematical Sciences: A Report* (Washington, DC: National Academies Press, 1968), 49.

16. Steven Shapin, "The Ivory Tower: The History of a Figure of Speech and Its Cultural Uses," *British Journal for the History of Science* 45, no. 1 (2012): 1–27.

17. David Kaiser and Benjamin Wilson, "Calculating Times: Radar, Ballistic Missiles, and Einstein's Relativity," in *Nation and Knowledge: Science and Technology in the Global Cold War*, ed. Naomi Oreskes and John Krige (Cambridge: MIT Press, 2014): 273–316.

18. Albers and Alexanderson, *Mathematical People*, 258.

19. Mina Rees, "Women Mathematicians Before 1950," *Newsletter of the AWN* 9 (August 1979): 15.

20. On Rees's work, see Amy Shell-Gellasch, *In Service to Mathematics: The Life and Work of Mina Rees* (Boston: Docent, 2011).

21. The story is recalled in Mina Rees, "Mathematics and the Government: Post-War Years as Augury of the Future," in *The Bicentennial Tribute to American Mathematics 1776–1976*, ed. Dalton Tarwater (Washington, DC: Mathematical Association of America, 1977), 101–16.

22. The American Mathematical Society has recognized Rees's contribution, noting that the fact that mathematical research did not lag behind in the aftermath of WWII "is beyond any doubt traceable to one person—Mina Rees. Under her guidance, basic research in general, and especially in mathematics, received the most intelligent and whole-hearted support. No greater wisdom and foresight could have been displayed and the whole postwar development of mathematical research in the United States owes in immeasurable debt to the pioneer work of the office

of Naval Research and to the alert, vigorous and farsighted policy conducted by Miss Rees." See L. W. Cohen, "The Annual Meeting in Baltimore," *Bulletin of the American Mathematical Society* 60, no. 2 (1954): 134.

23. Mina Rees, "The Mathematics Program of the Office of Naval Research," *Bulletin of the American Mathematical Society* 54, no. 1 (1948): 1–5.

24. Rees, "Mathcmatics Program."

25. Mina Rees, "Mathematics and Federal Support," *Science* 119 (1954): 3.

26. William L. Duren Jr., "The Support of Mathematical Research by the National Science Foundation," *Bulletin of the American Mathematical Society* 59, no. 1 (1953): 1. Emphasis in original.

27. Mark Solovey, *Shaky Foundations: The Politics-Patronage-Social Science Nexus in Cold War America* (New Brunswick, NJ: Rutgers University Press, 2013), 12.

28. Adrian Albert, "The Support of Mathematics at the National Science Foundation," November 13, 1952, AAAP, box 1, folder "A. A. Albert—Personal."

29. Albert, "Support of Mathematics."

30. Albert went even further in reaffirming the supposed hierarchy between pure and applied mathematics. He wrote that while he fully appreciated the importance of fields such as mathematical statistics and fluid mechanics, these fields were receiving "an emphasis with respect to support which [was] entirely unjustified by their importance relative to that of basic mathematics" because "such fields [were] more readily understood by the lay mind." Albert, "Support of Mathematics."

31. Adrian Albert, "The Needs of American Mathematics," undated, AAAP, box 1, folder "A. A. Albert—Personal." My emphasis.

32. Vannevar Bush, *Science: The Endless Frontier; A Report to the President* (Washington, DC: National Science Foundation, 1945), 6.

33. Richard Courant, "Diary Notice," November 29, 1943, RAMP, box 6, folder "Courant Diary."

34. Richard Courant, "Diary Notice for W. W. Concerning Conversation with H. Weyl," January 8, 1944, RAMP, box 6, folder "Courant Diary."

35. As noted in chapter 3, in the immediate aftermath of the war, what constituted applied mathematics was ambiguous. Some mathematicians advocated a big-tent approach that would include newer mathematical subfields such as operations research and communication theory, but for mathematicians like Albert, applied mathematics did not equal mathematization. Very crudely, applied mathematics denoted research that was oriented toward constructing new mathematical theories, and not economic, engineering, or physical theories.

36. "Current Activities of A. A. Albert," December 1954, AAAP, box 1, folder "A. A. Albert—Personal."

37. Saunders Mac Lane to Leon Cohen, July 1954, MMP, box 13, folder "Stone."

38. Leon Cohen to Joachim Weyl, June 21, 1954, AMSR, box 39, folder 50.

39. Adrian Albert to Members of the Division of Mathematics, undated, AMSR, box 40, folder 4.

40. Marston Morse, "Comments on a Survey of Training and Research in Applied Mathematics," June 8, 1954, MMP, box 13, folder "Stone."

41. Marshall Stone to Detlev Bronk and William Rubey, June 27, 1954, MMP, box 13, folder "Stone."

42. Marston Morse to Alan Waterman, Mina Rees, T. J. Killian, and Colonel C. G. Haywood, April 10, 1952, JWTP, series I, box 24, folder "National Research Council—Committee on Training and Research in Applied Mathematics."

43. "Applied Mathematics in the Scientific Community," AMSR, box 40, folder 4.

44. The AMS was by far the largest professional organization for research mathematicians at the time. The Mathematical Association of America was historically concerned with mathematical education. Thus, it was clear that the report was here directly referring to the AMS.

45. "Applied Mathematics."

46. Richard Courant to Joachim Weyl, January 5, 1954, RCP, box 24 (old), folder "Weyl F. J."

47. Constance Reid, *Courant* (New York: Springer, 1996).

48. Richard Courant and Herbert Robbins, *What Is Mathematics? An Elementary Approach to Ideas and Methods* (New York: Oxford University Press, 1996), 1.

49. See "General Statement–Brief History," RCP, box 45 (new), folder 7.

50. Even after adjusting for inflation, the annual budget increased by more than a factor of ten. More than half of the institute's budget bankrolled the installation of the UNIVAC at NYU, a project subsidized by the Atomic Energy Commission. Even if one ignores the cost of UNIVAC, the institute's budget increased by more than a factor of five. "General Statement–Brief History."

51. "Institute of Mathematical Sciences New York University: Address of R. Courant to Meeting of Advisory Board of the Institute of Mathematical Sciences," May 13, 1953, RCP, box 22 (new), folder 2.

52. "Institute of Mathematical Sciences New York University."

53. Ingram Olkin, "A Conversation with Albert H. Bowker," *Statistical Science* 2, no. 4 (1987): 477.

54. Olkin, "Albert H. Bowker," 476.

55. Richard Courant to Members of the Committee of Research and Training in Applied Mathematics, June 24, 1954, RCP, box 16 (old), folder "NRC."

56. "Summary and Recommendations," RCP, box 16 (old), folder "NRC."

57. "Summary and Recommendations."

58. "Summary and Recommendations."

59. Since the report criticized the AMS, and since he was the only pure mathematician on the committee, Stone felt under attack. In June, he wrote to the other members of the committee protesting the criticism directed at him by John Tukey. "There is no point in writing about personalities at this time," Stone wrote, "though I certainly feel entitled to characterize some of what he has written as innuendo unworthy of a respected colleague." Stone argued that the idea for the survey was his, and

hence it was unfair to suggest that he was hostile to the growth of the field. Tensions ran high on both sides. Marshall Stone to H. W. Bode, R. Courant, E. J. McShane, A. H. Taub, and J. W. Tukey, June 18, 1954, MMP, box 13, folder "Stone."

60. Marshall Stone, "Committee on Training and Research in Applied Mathematics: Minority Report," MMP, box 13, folder "Stone."

61. Stone, Committee on Training."

62. Samuel Eilenberg later made the case in an essay on the algebraization of mathematics addressed to a general readership. Eilenberg begins by explaining the general concept of the group as "an example of the general mental process of abstraction," and notes the growing use of algebraic methods across various mathematical subfields. He then adds that the way in which algebra impacts other parts of mathematics "has been compared with the relation of mathematics to other sciences." The mathematical theory of fluid dynamics, he explains, might consist of a model of the system that "disregards a large number of important properties of fluids . . . But because it is so oversimplified, we can treat it by mathematical methods, and because it mirrors correctly some physical reality, it permits us to draw important conclusions." Algebraization works in just the same manner, in that it involves "a drastic oversimplification of the mathematical concepts we study, but by concentrating on only aspects of mathematical reality, we are able to use our knowledge to study other parts of mathematics." Samuel Eilenberg, "The Algebraization of Mathematics," in *The Mathematical Sciences: A Collection of Essays*, ed. National Research Council, Committee on Support of Research in the Mathematical Sciences (Cambridge, MA: MIT Press, 1969), 153–60.

63. Eugene P. Wigner, "The Unreasonable Effectiveness of Mathematics in the Natural Sciences," *Communications on Pure and Applied Mathematics* 13, no. 1 (1960): 2.

64. Wigner, "Unreasonable Effectiveness of Mathematics."

65. Hans Hahn, "Logic, Mathematics, and Knowledge of Nature," in *Logical Positivism*, ed. A. J. Ayer (New York: Free Press, 1959), 151.

66. Hahn, "Logic, Mathematics, and Knowledge," 151.

67. Wigner, "Unreasonable Effectiveness of Mathematics," 14.

68. See, for example, the essays in the following books: Ronald N. Giere and Alan W. Richardson, ed., *Origins of Logical Empiricism* (Minneapolis: University of Minnesota Press, 1996); Alan W. Richardson and Gary L. Hardcastle, ed., *Logical Empiricism in North America* (Minneapolis: University of Minnesota Press, 2003); Alan Richardson and Thomas Uebel, ed., *The Cambridge Companion to Logical Empiricism* (Cambridge: Cambridge University Press, 2007).

69. Richard W. Hamming, "Numerical Analysis vs. Mathematics," *Science* 148, no. 3669 (April 23, 1965): 473–75.

70. Adrian A. Albert et al., "Mathematics vs. Numerical Analysis," *Science* 149, no. 3681 (July 16, 1965): 243–45.

71. Hamming, "Numerical Analysis vs. Mathematics," 474.

72. Albert et al., "Mathematics vs. Numerical Analysis," 243–44.

73. Albert et al., 244.

74. Albert et al.

75. Marshall H. Stone, "The Revolution in Mathematics," *Liberal Education* 47, no. 2 (May 1961): 305.

76. Stone, "Revolution in Mathematics," 307.

77. Stone, "Revolution in Mathematics," 310.

78. Stone, "Revolution in Mathematics."

79. Among others, the mathematicians who took part in the work of the committee included Mina Rees, William Prager, Allen Newell, Harold Grad, Joseph LaSalle, Andrew Gleason, and Adrian Albert.

80. National Research Council, *Mathematical Sciences*, 49.

81. National Research Council, 3.

82. National Research Council, 46.

83. National Research Council, 46.

84. Mina Rees, "The Nature of Mathematics," *Science* 138, no. 3536 (1962): 12.

85. Stanislaw M. Ulam, "The Applicability of Mathematics," in National Research Council, *Mathematical Sciences*, 1.

86. Ulam, "Applicability of Mathematics," 2.

87. Thomas L. Saaty and F. Joachim Weyl, *The Spirit and the Uses of the Mathematical Sciences* (New York: McGraw-Hill, 1969), 1.

88. Saaty and Weyl, *Spirit and the Uses*, 1.

89. Wigner, "Unreasonable Effectiveness of Mathematics."

90. George F. Carrier et al., "Applied Mathematics: What Is Needed in Research and Education: A Symposium," *SIAM Review* 4, no. 4 (1962): 298.

91. Carrier et al., "Applied Mathematics," 298.

92. Carrier et al., 298.

93. Richard Courant, "On the Graduate Study of Mathematics," in Weyl, *Proceedings of a Conference*, 30–34.

94. A department offered a specialization in a given field when its faculty included a professor who was able to direct a dissertation in the specific area of research as well as offer courses.

95. Mathematical statistics fared similarly. It had already emerged as a professional identity in the 1930s, but in the 1950s and 1960s statistics departments began appearing in universities across the country. By 1967 there were about thirty-one independent departments of statistics (at the time there were 126 mathematics departments) in the United States, and the number of PhDs granted in the field doubled from 67 in 1961 to 132 in 1968. Alan Agresti and Xiao-Li Meng, *Strength in Numbers: The Rising of Academic Statistics Departments in the U.S.* (New York: Springer, 2012).

96. H. J. Greenberg and Gareth Williams, ed., "Proceedings of a Conference on Education in Applied Mathematics," *SIAM Review* 9, no. 2 (April 1967): 320–21.

97. Greenberg, "Education in Applied Mathematics," 322.

98. Greenberg, "Education in Applied Mathematics."

99. It is important to emphasize that Bers's defense of abstract mathematics was not merely a cynical means of ensuring support for the field. More likely, it was a philosophical conviction. For one thing, Bers himself was an applied mathematician. More than that, Bers was aware of the need to develop and teach mathematics for practical purposes. Before he began reading his prepared remarks, Bers chided some of the mathematicians in attendance:

> I am afraid that during yesterday's discussion there appeared a certain attitude of snobbishness toward technologists, toward people who merely apply mathematics, rather than create new models and methods. I think this is a wrong attitude. Let us not fool ourselves. Society holds mathematics in high regard, not perhaps as it should because of the intrinsic beauty of the subject, but because mathematics is a useful art which society needs very badly. Our responsibility is not only to train the future leaders of thought in applied mathematics, but also to teach all people who will do the work of society. (Greenberg, "Education in Applied Mathematics," 315)

100. Greenberg, "Education in Applied Mathematics," 316.

101. Greenberg, 347.

102. Greenberg, 372.

103. Greenberg, 372.

104. Greenberg, 323.

105. Mark Kac, *Statistical Independence in Probability, Analysis and Number Theory* (Washington, DC: Mathematical Association of America, 1959), x.

106. Kac, *Statistical Independence*, 24.

107. Greenberg, "Education in Applied Mathematics," 315.

108. Morris Kline, *Mathematics in Western Culture* (Oxford, Oxford University Press, 1964); Morris Kline, *Mathematics in the Modern World: Readings from Scientific American* (New York: W. H. Freeman, 1968); Morris Kline, *Mathematics and the Physical World* (London: John Murray, 1959); Morris Kline, *Mathematical Thought from Ancient to Modern Times*, vol. 2 (New York: Oxford University Press, 1990); Morris Kline, *Mathematical Thought from Ancient to Modern Times*, vol. 3 (New York: Oxford University Press, 1990).

109. Kline, *Mathematical Thought*, vol. 3, 1036.

Chapter Six

1. Gottlob Frege, *Collected Papers on Mathematics, Logic, and Philosophy*, ed. Brian McGuinness (New York: Oxford University Press, 1984), 56.

2. Gottlob Frege, *The Foundations of Arithmetic: A Logico-Mathematical Enquiry into the Concept of Number*, trans. J. L. Austin (Evanston, IL: Northwestern University Press, 1980).

3. Michael Dummett, *Frege: Philosophy of Language* (Cambridge, MA: Harvard University Press, 1981), 2.

4. Herbert Mehrtens, "Mathematical Models," in *Models: The Third Dimension of Science*, ed. Nick Hopwood and Soraya de Chadarevian (Stanford, CA: Stanford University Press, 2004), 300.

5. Mehrtens, "Mathematical Models."

6. The historiography of mathematics in the postwar period can be divided in two. During the 1950s and 1960s, the subject was almost the sole purview of historically minded mathematicians. Popular books offering *long durée* histories of mathematics for a general readership and mathematical students were plenty. Dirk Jan Struik's *A Concise History of Mathematics* (1948), Edna E. Krammer's *The Main Stream of Mathematics* (1951), Howard Eves's *An Introduction to the History of Mathematics* (1953), Morris Kline's *Mathematics in Western Culture* (1953), Otto Neugebauer's *The Exact Sciences in Antiquity* (1957), J. F. Scott's *A History of Mathematics: From Antiquity to the Beginning of the 19th Century* (1958), Salomon Bochner's *The Role of Mathematics in the Rise of Science* (1966), and Carl B. Boyer's *A History of Mathematics* (1968) are only a few of the titles published in English during the first two and a half decades after World War II. Many other books on the history of mathematics were translated at the time from other languages. For example, B. L. van der Waerden's book *Science Awakening* was translated and published in English in 1961. Otto Neugebauer's work stands out in this list. Neugebauer was an expert in the history of ancient mathematics and astronomy, and his work had profound influence on the study of ancient mathematics. Alexander Jones, Christine Proust, and John M. Steele, *A Mathematician's Journeys: Otto Neugebauer and Modern Transformations of Ancient Science* (New York: Springer, 2016).

7. Quentin Skinner, "Meaning and Understanding in the History of Ideas," *History and Theory* 8, no. 1 (1969): 3–53.

8. Joel Isaac, *Working Knowledge: Making the Human Sciences from Parsons to Kuhn* (Cambridge, MA: Harvard University Press, 2012), 216.

9. Peter Galison, *Image and Logic: A Material Culture of Microphysics* (Chicago: University of Chicago Press, 1997), 791.

10. Thomas Kuhn, "Reflections on My Critics," in *Criticism and the Growth of Knowledge*, ed. Imre Lakatos and Alan Musgrave (Cambridge: Cambridge University Press, 1970), 266.

11. Ludwig Wittgenstein, *Philosophical Investigations*, trans. G. E. M. Anscombe (New York: Macmillan, 1953), 20. My emphasis.

12. Skinner remarks on the influence of Wittgenstein and J. L. Austin on his thinking in several interviews. See, for example, Raia Prolhovnik, "An Interview with Quentin Skinner," *Contemporary Political Theory* 10, no. 2 (2011): 273–85.

13. For a defense of the Platonist position, see Jean-Pierre Changeux and Alain Connes, *Conversations on Mind, Matter, and Mathematics* (Princeton, NJ: Princeton University Press, 1998).

14. Elaine Koppelman, "Progress in Mathematics," *Historia Mathematica* 2, no. 4 (1975): 462.

15. Koppelman, "Progress in Mathcmatics," 463.

16. Koppelman.

17. Koppelman, 463.

18. See, for example, Susan M. Reverby and David Rosner, "'Beyond the Great Doctors' Revisited: A Generation of the 'New' Social History of Medicine," in *Locating Medical History: The Stories and Their Meanings*, ed. John Harley Warner and Frank Huisman (Baltimore: John Hopkins University Press, 2004), 167–93.

19. In the history of mathematics, however, these debates continued well into the 1990s. For example, in 1995, Joan Richards reflected that while the internalist-externalist debate had been declared dead in the history of science, "the division between the two camps [was] not only *a* but *the* critical problem in the history of mathematics." Joan L. Richards, "The History of Mathematics and *L'esprit Humain*: A Critical Reappraisal," *Osiris* 10, Constructing Knowledge in the History of Science (1995): 123–24.

20. Koppelman, "Progress in Mathematics," 463.

21. Thomas S. Kuhn, "Comment on the Relations of Science and Art," in *The Essential Tension: Selected Studies in Scientific Tradition and Change* (Chicago: University of Chicago Press, 1977), 345.

22. Kuhn, "Science and Art," 345.

23. Koppelman, "Progress in Mathematics."

24. Michael Crowe, "Ten 'Laws' Concerning Conceptual Change in Mathematics," *Historia Mathematica* 2, no. 4 (1975): 470.

25. Crowe, "Ten 'Laws,'" 470.

26. Crowe, 470.

27. Herbert Mehrtens, "T. S. Kuhn's Theories and Mathematics: A Discussion Paper on the 'New Historiography' of Mathematics," *Historia Mathematica* 3, no. 3 (1976): 301.

28. Mehrtens also made another important interjection about the use of the preposition "in." As he explained, in adopting the preposition, historians of mathematics inevitably followed mathematicians' own definition of what was and was not mathematics. For a mathematician, as Mehrtens explained, the historical record that counted as being "in" mathematics was only that which still seemed to speak to the present: "Everything which is included in or derivable from modern mathematics is *in* mathematics. The historically significant features like use of concepts, the general beliefs concerning the discipline, etc. are naturally not *in* mathematics." Mehrtens, "T. S. Kuhn's Theories," 302.

29. Thomas Hawkins, "Mathematical Progress without Fusion," *Historia Mathematica* 2, no. 4 (1975): 565.

30. Hawkins, "Mathematical Progress without Fusion," 565–66.

31. Hawkins, 566.

32. Hawkins, 566.

33. Kuhn, "Reflections on My Critics," 236.

34. This is not to suggest that Kuhn's thinking was axiomatic. In a paper reflecting on J. D. Sneed's *The Logical Structure of Mathematical Physics*, Kuhn clearly writes, "Traditional formalisms, whether set-theoretical or propositional, have made no contact whatsoever with mine." Thomas S. Kuhn, "Theory-Change as Structure-Change: Comments on the Sneed Formalism," *Erkenntnis* 10, no. 2 (1976): 179. It is worth noting, however, that Kuhn spent 1958–1959 as a fellow at the Center for Advanced Study in the Behavioral Sciences at Stanford University. Among the other fellows in the center that year were Clifford Geertz, Roman Jakobson, George A. Miller, and Willard V. Quine.

35. Peter Gordon, "Forum: Kuhn's Structure at Fifty Introduction," *Modern Intellectual History* 9, no. 1 (2012): 74.

36. Claude Lévi-Strauss, "The Mathematics of Man," *International Social Science Bulletin* 6, no. 4 (1954): 581–90.

37. Gordon, "Kuhn's Structure at Fifty," 75.

38. Gordon, 75.

39. Anthony Grafton, "The History of Ideas: Precept and Practice, 1950–2000 and Beyond," *Journal of the History of Ideas* 67, no. 1 (2006): 1–32.

40. Galison, *Image and Logic*, 796.

41. Joel Isaac, "Kuhn's Education: Wittgenstein, Pedagogy, and the Road to Structure," *Modern Intellectual History* 9, no. 1 (2012): 89–107.

42. Thomas S. Kuhn, *The Structure of Scientific Revolutions* (Chicago: University of Chicago Press, 1970), 44–45.

43. Kuhn, *Structure of Scientific Revolutions*, 45.

44. Anatol Rapoport, "Various Meanings of 'Theory,'" *American Political Science Review* 52, no. 4 (1958): 979.

45. Skinner, "Meaning and Understanding," 37.

46. Skinner, 9.

47. Charles V. Jones, Philip C. Enros, and Henry S. Tropp, "Kenneth O. May, 1915–1977: His Early Life to 1946," *Historia Mathematica* 11, no. 4 (1984): 359–79.

48. In 1967, for example, May published "Mathematics and Art" in *The Mathematics Teacher*. Like many of his colleagues described in chapter 5, May celebrated the bond between mathematics and art. "I maintain, then," May wrote, "that the revolutions in art and mathematics have only deepened the relations between them and this is borne out by the psychology of practitioners in both fields." He added, "It is a common observation that the emotional drive for creation and the satisfaction from success are the same whether one is constructing an art object

or a mathematical theory." Kenneth O. May, "Mathematics and Art," *Mathematics Teacher* 60, no. 6 (1967): 572.

49. The publication of the bibliography indicates the growing interest in the history of mathematics. However, May's effort was not the first. In 1952, Cecil B. Read of the University of Wichita compiled a more restricted bibliography of all the articles on the history of mathematics, which appeared in five chosen periodicals. Kenneth O. May, *Bibliography and Research Manual of the History of Mathematics* (Toronto: University of Toronto Press, 1973); Albert C. Lewis, "Kenneth O. May and Information Retrieval in Mathematics," *Historia Mathematica* 31, no. 2 (2004): 186–95. Cecil B. Read, "Articles on the History of Mathematics: A Bibliography of Articles Appearing in Five Periodicals," *Municipal University of Wichita Bulletin* 26, no. 4 (1952).

50. It was not, however, the first journal dedicated to the history of mathematics. In 1932, Jekuthiel Ginsberg helped found *Scripta Mathematica*, which was dedicated to the history and philosophy of mathematics (it ceased publication in 1973). Another journal dedicated to research in the history of mathematics was *Archive for History of Exact Sciences*. Unlike the popular histories of mathematics written in the 1950s and 1960s, these journals were dedicated to scholarly research articles in the history of mathematics.

51. Kenneth O. May, "Should We Be Mathematicians, Historians of Science, Historians, or Generalists?" *Historia Mathematica* 1, no. 2 (May 1974): 127–28.

52. Kenneth O. May, "Historiographic Vices I. Logical Attribution," *Historia Mathematica* 2 (1975): 185–87.

53. To illustrate the problem, May retells a common anecdote regarding Lefschetz: "The hero is lecturing to a seminar. 'Now the following result is obvious. Hmmm. Or is it? Hmmm. Excuse me, I'll be back in a moment.' He leaves. Twenty minutes pass. He returns. 'Yes, it's obvious!'" May, "Should We Be Mathematicians," 186.

54. Kenneth O. May, "Historiographic Vices II. Priority Chasing," *Historia Mathematica* 2 (1975): 315–17.

55. May, "Logical Attribution," 187.

56. Unguru's work has received much scholarly attention. Most recently, Martina Schneider has argued that the controversy must be understood in the context of the professionalization of the history of mathematics in the mid-1970s. While I certainly agree with Schneider, I want to look at the controversy in order to unpack the different conceptions of mathematics that undergirded the debate. Martina R. Schneider, "Contextualizing Unguru's 1975 Attack on the Historiography of Ancient Greek Mathematics," in *Historiography of Mathematics in the 19th and 20th Centuries*, ed. Volker R. Remmert, Martina R. Schneider, and Henrik Kragh Sørensen (Cham, Switzerland: Birkhäuser, 2016), 245–68.

57. Sabetai Unguru, "On the Need to Rewrite the History of Greek Mathematics," *Archive for History of Exact Sciences* 15, no. 1 (1975): 68. Emphasis in original.

58. Unguru, "On the Need," 69.

59. It is worth noting that Unguru makes it clear that he was not the first to voice such a critique and repeatedly refers in his text and footnotes to articles by Michael Mahoney.

60. Unguru, "On the Need," 86.

61. Skinner, "Meaning and Understanding," 10.

62. Skinner, 22.

63. Hans Freudenthal, "What Is Algebra and What Has It Been in History?" *Archive for History of Exact Sciences* 16, no. 3 (1977): 190.

64. Sabetai Unguru, "History of Ancient Mathematics: Some Reflections on the State of the Art," *Isis* 70, no. 4 (1979): 555–56.

65. Unguru, "History of Ancient Mathematics," 562.

66. Unguru, 563.

67. I do not mean to imply here that Unguru is wrong. By his definition of history, neither mathematics nor the work of mathematicians qualifies as history. Mathematicians' claiming mathematics as history was a reflection on what mathematics was as a body of knowledge and as an activity.

68. Clifford Truesdell, *Essays on the History of Mechanics* (New York: Springer-Verlag, 1968), i.

69. Truesdell, *History of Mechanics*, i.

70. One example of this is the Mathematics Genealogy Project, a website that tracks the PhD advisor of every mathematician. The website offers personalized genealogy posters as "a great way to display your academic heritage." Such a genealogy poster does not simply go back two or three generations but stretches over centuries. For example, the genealogy tree of a recent graduate from the University of Wisconsin–Madison begins with Georgios Gemistos Plethon, the fourteenth-century philosopher. Such genealogy trees are only one way in which mathematicians conceive of their work historically. Mathematics Genealogy Project, https://genealogy.math.ndsu.nodak.edu.

71. John Archibald Wheeler, "Hermann Weyl and the Unity of Knowledge," *American Scientist* 74, no. 4 (August 1986): 366–75. My emphasis.

72. Morris Kline, "Opening Remarks," *Historia Mathematica* 2, no. 4 (1975): 575.

73. André Weil, "History of Mathematics: Why and How," in *Proceedings of the International Congress of Mathematicians*, vol. 1 (Helsinki: Academia Scientiarum Fennica, 1980), 229.

74. Weil, "History of Mathematics," 229.

75. David E. Rowe, "New Trends and Old Images in the History of Mathematics," in *Vita Mathematica: Historical Research and Integration with Teaching*, ed. Ronald Calinger (Washington, DC: Mathematical Association of America, 1996), 10.

76. Weil, "History of Mathematics," 230.

77. Weil, 235–36.

78. Weil, 236. My emphasis. Ivor Grattan-Guinness wrote that mathematicians' approach to history can often be described as "a 'royal road to me'—that is, an

account of how a particular modern theory arose out of older theories instead of an account of those older theories in their own right." This view is limited, as I show. Mathematicians turned to history seeking inspiration for their current work. The goal of decoding the precursors for one's idea in the work of Carl Friedrich Gauss, or some other towering figure in the history of mathematics, is not to praise one's own achievement, but instead to show how one's work is part of a longer mathematical conversation. Ivor Grattan-Guinness, "Does History of Science Treat of the History of Science? The Case of Mathematics," *History of Science* 28, no. 2 (1990), 157.

79. In a 2004 edited volume that published Unguru's original paper and its responses, Unguru concluded, "It turns out that the mathematical and historical approaches are mutually antagonistic and that no compromise is possible between the mathematical and historical methodological principles. Adopting one or the other has fateful consequences for one's research, effectively determining the nature of the results reached and the tenor of the inferences used in reading them." Sabetai Unguru, "Introduction," in *Classics in the History of Greek Mathematics*, ed. Jean Christianidis, vol. 240, Boston Studies in the Philosophy of Science (Dordrecht, Neth.: Springer Science+Business Media Dordrecht, 2004), 383.

80. Skinner, "Meaning and Understanding," 50.

81. Skinner, 50. On the evolution of Skinner's thinking, see Melissa Lane, "Doing Our Own Thinking for Ourselves: On Quentin Skinner's Genealogical Turn," *Journal of the History of Ideas* 73, no. 1 (2012): 71–82.

82. Christopher J. Berry, "Kuhn and the History of Ideas," *British Journal for the History of Science* 12, no. 3 (1979): 297.

83. Armitage is clear to note that such *long durée* studies are already growing in popularity. David Armitage, "What's the Big Idea? Intellectual History and the Longue Durée," *History of European Ideas* 38, no. 4 (2012): 497.

84. Armitage, "What's the Big Idea?."

85. Martin Jay, "'Hey! What's the Big Idea?': Ruminations on the Question of Scale in Intellectual History," *New Literary History* 48, no. 4 (2017): 622.

86. Darrin M. McMahon, "The Return of the History of Ideas?," in *Rethinking Modern European Intellectual History*, ed. Darrin M. McMahon and Samuel Moyn (New York: Oxford University Press, 2014), 21.

87. McMahon, "The Return of the History of Ideas?," 22.

88. John Tresch, "Cosmologies Materialized: History of Science and History of Ideas," in *Rethinking Modern European Intellectual History*, ed. Darrin M. McMahon and Samuel Moyn (New York: Oxford University Press, 2014), 155.

89. Interestingly, both Martin Jay and historian of mathematics Moritz Epple have turned to music as an analogy with which to account for the timeless and historical nature of ideas. Jay writes that "history urges us to follow the musical model of themes and developing variations, without, of course, necessarily adopting the practice of tonal recapitulation fundamental to classical Western music." Epple, who adapts Hans-Jörg Rheinberger's notion of an "epistemic thing" from

laboratory studies to mathematics, insists on the contextualized reading of mathematical texts. Epple argues that definitions only exist within technical frameworks and that the epistemic objects of mathematics are "secondary to the dynamics of the epistemic configurations as a whole." Mathematical notions cannot be studied in the abstract, but only "as they present and [are] involved in actual research." Yet he too recognizes that this approach does not solve the philosophical question regarding the temporality of mathematical things, which leads him to ask whether the "temporal character of mathematical definitions (and even of full mathematical arguments) have some similarity with that of musical compositions." There is no avoiding the philosophical. Moritz Epple, "Between Timelessness and Historiality: On the Dynamics of the Epistemic Objects of Mathematics," *Isis* 102, no. 3 (2011), 492.

90. Quoted in Jay, "'What's the Big Idea?'" 622.

91. Jay, "'What's the Big Idea?'" 623.

92. As Skinner writes, even if we only subscribe to a Wittgensteinian notion of "family resemblance . . . we are still committed to accepting *some* criteria and rules of usage such that certain performances can be correctly instanced, and others excluded, as examples of a given activity. Otherwise we should eventually have no means—let alone justification—for delineating and speaking, say, of the histories of ethical or political thinking as being histories of recognizable activities at all." Moreover, Skinner writes, it is the "truth" rather than the "absurdity" of the claim that moral and political thinking is concerned with some "characteristic concepts" that is the source of much "confusion." Skinner, "Meaning and Understanding," 6.

93. Skinner, "Retrospect: Studying Rhetoric and Conceptual Change," in *Visions of Politics* (Cambridge, Cambridge University Press, 2002), 177.

94. Skinner, "Retrospect," 178.

95. Skinner, "Retrospect," 179.

96. Skinner, 186.

97. Philip J. Davis and Reuben Hersh, *The Mathematical Experience* (Boston: Houghton Mifflin Harcourt, 1981).

98. Similarly, Wickberg argues that Lovejoy's notion of the "unit-idea," which has been the target of much criticism, was not meant as an ontological claim. Rather, it was a methodological claim, "a way to examine bodies of thought in order to identify both similarities and differences over time and between texts, a way to examine contradictions and opposing outcomes produced by the tension involved in what he called idea complexes." Daniel Wickberg, "In the Environment of Ideas: Arthur Lovejoy and the History of Ideas as a Form of Cultural History," *Modern Intellectual History* 11, no. 2 (2014): 448.

99. Wickberg, "Environment of Ideas," 454.

100. Friedrich Nietzsche, *On The Genealogy of Morals*, ed. Walter Kaufmann (New York: Vintage, 1989), 80.

Epilogue

1. George K. Francis, interview with author, June 3, 2011.

2. George K. Francis, *A Topological Picturebook* (New York: Springer, 2006).

3. George K. Francis, interview with author, June 2, 2011.

4. William Abikoff, *The Real Analytic Theory of Teichmüller Space* (New York: Springer-Verlag, 1980); William Abikoff, "The Uniformization Theorem," *American Mathematical Monthly* 88, no. 8 (1981): 574–92.

5. The notes from Thurston's 1978 course on the geometry and topology of three-dimensional manifolds were sent to over a thousand mathematicians around the world. Athanase Papadopoulos wrote, "It is probably the opinion of all the people working in low-dimensional topology that the ideas contained in these notes have been the most important and influential ideas ever written on the subject." Athanase Papadopoulos, "Review: Three-Dimensional Geometry and Topology, Vol. 1," *Mathematical Review* (1997), 1435975.

6. Papadopoulos, "Three-Dimensional Geometry."

7. William P. Thurston, "How to See 3-Manifolds," *Classical and Quantum Gravity* 15, no. 9 (1998): 2545–71.

8. William P. Thurston, "Three-Dimensional Manifolds, Kleinian Groups and Hyperbolic Geometry," *Bulletin of the American Mathematical Society* 6, no. 3 (1982): 357–81; William P. Thurston, "Hyperbolic Structures on 3-Manifolds. I: Deformation of Acylindrical Manifolds," *Annals of Mathematics*, Second Series 124, no. 2 (1986): 203–46; William P. Thurston, *Three-Dimensional Geometry and Topology*, ed. Silvio Levy (Princeton, NJ: Princeton University Press, 1997).

9. Felix E. Browder, "The Relation of Functional Analysis to Concrete Analysis in 20th Century Mathematics," *Historia Mathematica* 2, no. 4 (1975): 577–78.

10. Here it is important to emphasize that I do not claim that abstract and axiomatic mathematics have disappeared from mathematical research. Quite the contrary. Many of the developments of the 1970s and 1980s could easily qualify as part of that tradition. Rather, I wish to note that the abstract approach to mathematical research that always placed the general above the concrete was no longer the *only* legitimate approach to mathematics.

11. Siobhan Roberts, *King of Infinite Space: Donald Coxeter, the Man Who Saved Geometry* (New York: Walker & Co., 2006).

12. Jean Dieudonné, "The Universal Domination of Geometry," *Two-Year College Mathematics Journal* 12, no. 4 (1981): 231.

13. Branko Grünbaum, "Shouldn't We Teach Geometry?" *Two-Year College Mathematics Journal* 12, no. 4 (1981): 232.

14. Robert Osserman, "Structure vs. Substance: The Fall and Rise of Geometry," *Two-Year College Mathematics Journal* 12, no. 4 (1981): 239. Branko Grünbaum, a geometer teaching at the University of Washington in Seattle, was also quick to criticize Dieudonné's position. "I cannot find much solace or joy in this view," he

retorted. What Dieudonné called geometry and what Grünbaum acknowledged as geometry were two separate things. Grünbaum, "Shouldn't We Teach Geometry?"

15. Osserman referred to this approach to mathematics as the "Bourbaki approach," but he was quick to add, "Bourbaki is undoubtedly a follower as well as a leader of the general trend. Also, whether Bourbaki specifically espouses that approach, he or she clearly epitomizes it." Osserman, "Structure vs. Substance," 239.

16. Osserman, "Structure vs. Substance," 243.

17. Osserman, 244.

18. The members of the group were Fredrick Almgren, James Cannon, B. Chaselle, John Conway, David Dobkin, Adrien Douady, David Epstein, Michael Freedman, Pat Hanrahan, John Hubbard, Harvey Keynes, Benoit Mandelbrot, Al Marden, John Milnor, David Mumford, Charles Peskin, William Thurston, and Allan Wilks.

19. I. Peterson, "Shareware, Mathematics Style," *Science News* 133, no. 1 (1988): 12–13.

20. Alma Steingart, "A Four-Dimensional Cinema: Computer Graphics, Higher Dimensions, and the Geometrical Imagination," in *Visualization in the Age of Computerization*, ed. Annamaria Carusi, Aud Sissel-Hoel, Timothy Webmoor, and Steve Woolgar (New York: Routledge, 2014), 170–96.

21. Thomas F. Banchoff, "Computer Graphics Application in Geometry: 'Because the Light Is Better Over Here,'" in *The Merging of Disciplines: New Directions in Pure, Applied, and Computational Mathematics*, ed. Richard Ewing, Kenneth Gross, and Clyde Martin (New York: Springer-Verlag, 1985), 1–14.

22. Osserman, "Structure vs. Substance," 241.

23. Osserman, 241.

24. Osserman, 242.

Index